Living in Groups

Oxford Series in Ecology and Evolution
Edited by Paul H. Harvey and Robert M. May

Living in Groups

JENS KRAUSE
University of Leeds

GRAEME D. RUXTON
University of Glasgow

OXFORD
UNIVERSITY PRESS

OXFORD
UNIVERSITY PRESS

Great Clarendon Street, Oxford OX2 6DP

Oxford University Press is a department of the University of Oxford.
It furthers the University's objective of excellence in research, scholarship,
and education by publishing worldwide in

Oxford New York

Auckland Bangkok Buenos Aires Cape Town Chennai
Dar es Salaam Delhi Hong Kong Istanbul Karachi Kolkata
Kuala Lumpur Madrid Melbourne Mexico City Mumbai Nairobi
São Paulo Shanghai Taipei Tokyo Toronto

Oxford is a registered trade mark of Oxford University Press
in the UK and in certain other countries

Published in the United States
by Oxford University Press Inc., New York

A catalogue record for this title is available from the British Library

Library of Congress Cataloging in Publication Data

ISBN 0 19 850817 4(hbk.)
ISBN 0 19 850818 2(pbk.)

10 9 8 7 6 5 4 3 2 1

Typeset by Newgen Imaging Systems (P) Ltd, Chennai, India
Printed in Great Britain
on acid-free paper by Biddles Ltd, Guildford & King's Lynn

To Josephine Wong and Hazel Ruxton

Preface

The idea for this book was first conceived in discussions with Ian Sherman (then at Blackwell Science) who referred us to the Oxford Series in Ecology and Evolution (OSEE) because it seemed the most suitable place for such a book. At that time we were thinking of writing a monograph on shoaling behaviour in fish which was meant to address many of the important issues concerning grouping behaviour. This idea formed the basis of our first book proposal which received mixed responses from both the reviewers and OUP. The main criticism was that a book on shoaling behaviour might be too specialised and would only be of interest to a relatively restricted audience. Kate Kilpatrick, OUP's Biology Editor, then gently persuaded us to broaden the scope of the book so that it would cover grouping animals in general. This seemed a rather daunting task at first given the enormous number of species from very diverse taxonomic groups that would have to be included. Also, the sheer limitless literature that would need to be reviewed presented serious obstacles. After some consideration we decided on the present format which is centred on the conceptual issues of grouping behaviour rather than on taxonomic diversity. We have made an effort to illustrate the concepts with examples from different taxonomic groups. However, our first priority was the quality of the studies and whether they suited our purpose. (Therefore this book is not meant to be an encyclopaedia of grouping animal species.) Nevertheless we have almost certainly overlooked many important studies and likewise it was not possible to present all arguments in their full length given the restricted space that was available. The result is a book which aims to give a broad overview of the literature on current theories and concepts with references to the more specialised texts wherever appropriate.

With a book as broad in scope as ours, it is necessary to make a few disclaimers and state explicitly what we do not cover and why. We do not explicitly discuss the evolution of cooperation, specifically by reciprocal altruism, in this book. One of the requirements for reciprocal altruism to work is that two individuals repeatedly interact (Trivers 1971). It is certainly true that being in a group of stable composition would facilitate such repeated interactions between two individuals. However it is not a prerequisite. The interested reader is directed towards a recent text in the same series devoted to cooperation in animals: Dugatkin (1997). Also, our treatment of the evolution of social insects is rather short to avoid overlap with the book by Crozier and Pamilo (1997) that was also published in the OSEE.

Despite our best efforts, there are probably some mistakes in this book; certainly there are some omissions. It was never our intention to provide a catalogue of all published works with a bearing on animal grouping, but there are probably some works that we overlooked but would have incorporated if only we had known about them. Hence, if readers spot any errors or feel that a particular work has been wrongly excluded, then we would be pleased to hear from you via e-mail: bgyjk@leeds.ac.uk or G.Ruxton@bio.gla.ac.uk. All errors and corrections brought to our attention will be posted on the OUP website at www.oup.co.uk/ISBN/0-19-850818-2.

Over the last two years a large number of people provided help and support in different forms while this project progressed from a book proposal to the final version.

First of all thanks are due to Ian Sherman whose visit to Leeds in 1999 started the whole project with his question: "Is there a book that you would like to write?" Ian was very supportive of our ideas and made the connection to the OSEE for us. It was under the guidance of Kate Kilpatrick, our initial OUP editor, that we made the transformation from a book on fish shoaling to one on grouping behaviour in general. Ian Sherman inherited this book at the stage where we had agreed to write it but had yet to actually write a word. We have both benefited greatly from discussions with Ian. He has provided a fascinating insight into publishing, as well as offering support and encouragement, strongly laced with good advice. Paul Harvey read every word of a full draft, and his perceptive comments spurred us on to make some (we hope) significant improvements.

We would like to acknowledge the people who provided particularly useful comments on specific areas of the text, although in no way do we wish to pass the blame for any errors or omissions on to them. Our thoughts on group selection were greatly improved by advice from Kevin Laland, Lukas Keller and Nick Colegrave. The sections on the benefits of foraging in a group in chapter 2 lean heavily on ongoing discussions between GDR, Rik Smith and Will Cresswell. Daniel Hoare and Iain Couzin made helpful comments on chapter 5. Chapter 6 was partly based on discussion with Roland Tegeder and greatly improved by perceptive feedback from Hazel Wright. Vincent Janik, Chris Thomas and Nick Davies provided us with critical feedback on chapter 7 and Charlotte Hemelrijk and Iain Couzin made helpful comments on chapter 9. We are particularly grateful to Steven Simpson for photos of his locusts and for his insightful comments on chapter 8.

Others provided us with useful references, specific ideas or simply with a quiet place to work: Katharina Riebel, Liz Hensor, Charlotte Hemelrijk and Stefan Krause. In particular, Hasso Neumann and Barbara Vogel provided ideal circumstances for a retreat at the Matteschlösschen where a substantial part of this book was written. Their support is gratefully acknowledged.

The smartest thing we did was use Liz Denton to design and draw the figures for us. She took awful photocopies and scribbles and turned them into the clear diagrams that grace this book. Not only that, but she is a model of quick and friendly efficiency that we can only aspire to. The sketches kindly provided by Vicky Mills further enhance the figures.

We have both benefited greatly from working in stimulating and interactive departments and would like to thank the many members of staff and students in both Glasgow and Leeds for providing such supportive and creative environments for scientific discussions to flourish.

Financial support was provided by the NERC, the Leverhulme Trust, and the Fisheries Society of the British Isles.

Contents

1

Introduction

1.1 Overview

Group-living is a widespread phenomenon in the animal kingdom and has attracted considerable attention in a number of different contexts. A large literature exists explaining the existence of grouping, concerning both its proximate and ultimate causes. Furthermore numerous studies on topics like self-organization, reciprocal altruism, and producer–scrounger relationships have used animal groups as study systems.

Given the vast literature on animal groups, we will not attempt to give a comprehensive review of all existing studies, as this would neither be possible nor desirable. Instead our focus will be on the unifying concepts regarding grouping behaviour that have been developed over the last two decades (e.g. the optimal group size concept; co-operative breeding). Our aim is to provide a discussion of the mechanisms that govern the evolution and maintenance of grouping behaviour throughout the animal kingdom, and the ecological factors that control observed group size and group composition in particular situations. Although the emphasis will be on the elaboration of a conceptual framework, extensive examples and case studies from as wide a range of taxonomic groups as possible will illustrate how widespread a phenomenon grouping is in the animal kingdom, and demonstrate the general applicability of the concepts developed.

The book will familiarize the reader with the current ideas on the ecology and evolution of group-living animals, providing a summary and critical synthesis of the extensive and diverse literature on this subject. Selected case studies will illustrate how these ideas and concepts are applied to actual study systems. Additionally the reader will be guided to the more specialized literature for further reading on certain topics that are beyond the scope of this book. For instance, social insect colonies (and eusocial animals in general for that matter) represent an important form of social organization and will be covered in Chapter 7. However, for details we will refer to the more specialized literature (e.g. Hölldobler and Wilson 1990; Crozier and Pamilo 1996) wherever appropriate. At the end of each chapter those aspects of both theory and empirical research most in need of further endeavour will be highlighted.

Our emphasis will be on the functions of grouping (Chapters 2–6), but developmental (Chapter 8), evolutionary (Chapter 7), and mechanistic aspects (Chapter 9) of group-living will be discussed as well. Thus we intend to cover all four aspects of grouping behaviour which Tinbergen (1963) identified as central to the study of animal behaviour in general. This means we will attempt to provide some answers to questions of why animals live in groups, what determines their size, shape, and composition, and how groups are formed and maintained.

1.2 A definition of groups

There are a number of different approaches to defining what an animal group is (Wilson 1975; Lee 1994; Pitcher and Parrish 1993). Wilson (1975) characterized a group as 'any set of organisms, belonging to the same species, that remain together for a period of time interacting with one another to a distinctly greater degree than with other conspecifics'. And Lee (1994) stated that 'when two or more animals live together they constitute a social unit'.

There seems to be general agreement that a certain degree of proximity in time and space is an essential prerequisite for grouping. Definitions such as Pitcher *et al.*'s (1983) elective group size concept assume that the inter-individual distance between group members is a function of the trade-off between the costs and benefits associated with group-living. Many animal species aggregate in the presence of a predation threat and disperse when foraging. Pitcher *et al.* (1983) observed that co-ordinated group behaviours were only possible in fish (for which this concept was originally developed) if the individuals stayed within at least four to five body lengths of each other. Thus the elective group size concept requires animals to be close enough for continuous information exchange between them. This distance can vary considerably between different species (e.g. cetaceans are capable of communicating acoustically over long distances) (Fig. 1.1). A further requirement for so-called social groups is that the animals are brought together by social attraction. This means that individuals actively seek the proximity of each other instead of co-occurring in the same spot because of an attraction to the same environmental condition or factor such as a localized food source or a rock for basking.

Searching for a perfect all-encompassing definition of grouping can easily result in a sterile list of criteria that are hard to apply and often rather arbitrary. Given the great diversity of animal grouping behaviour, it seems less important to find a definition that can be applied rigidly to all cases, rather than developing operational definitions of grouping that fit both the study species in question and more generally into the above framework of grouping.

We, nevertheless, caution researchers to take some time to explore the applicability of the various definitions to their study species because this can result in obtaining useful background information for almost any investigation. Collecting baseline data on the inter-individual distance between group members, the variation thereof, and the major communication channels employed for information exchange between individuals can be a useful starting point. Similarly, testing for social attraction by using a choice apparatus can be helpful (Fig. 1.2). Thus there is indeed a little list from which suitable criteria can be picked for almost any species of group-living

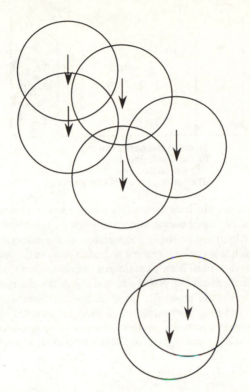

Fig. 1.1   Elective group size concept. On each individual (symbolized by an arrow) a circle is centred, the radius of which represents the maximum distance at which effective communication between individuals is possible resulting in synchronized behaviour. A circle is used because it is assumed that communication is purely a function of distance which should be equally possible in all directions. In practice, however, this is hardly ever true because of directional bias of sensory organs and environmental disturbances. Group size is given by the number of overlapping circles that are interconnected.

animal when setting out for an investigation of its behaviour in the laboratory or the field.

Another important point to consider is that the division of animal species into group-living and solitary ones is largely artificial. Many (if not most) species are intermediates that will be found in association with con- or heterospecifics at certain times but not always. This is even true of some species of which we generally think as always being in groups. When a frequency distribution of group sizes is recorded, it often becomes apparent that singletons (or very small groups) are the most frequently observed ones. This is a paradox to which we will return in Chapter 9 (Mechanisms of grouping). For the moment it is enough to accept that for many so-called group-living species the group size distribution can have a very wide range and usually will include many singletons. Therefore the purpose of this book is not to categorize species into those that group and those that don't, but to look at the

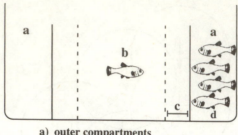

a) outer compartments
b) focal individual
c) association zone
d) stimulus group of conspecifics

Fig. 1.2 Choice apparatus which can be used for testing social attraction in animals. A focal individual is placed in the central compartment and given a choice between a group of conspecifics on one side (right) and an empty compartment on the other (left). The number and duration of visits to each side can be measured and taken as an indication of the attraction to conspecifics. The focal individual is only considered to have made a choice between the two outer compartments if it crosses the dotted line and enters the association zone representing the species-specific elective group size distance. Stimulus groups are presented equally often in the right and left outer compartment to control for potential side preferences. The inner walls of the outer compartments can be transparent or opaque and perforated or solid depending on whether or not visual and olfactory cues should be transmitted.

interactions that occur when animals come into close proximity (with each other) and the long- and short-term consequences of such interactions.

1.3 Book structure

In the following two chapters (Chapters 2 and 3), we will concentrate on the costs and benefits of grouping giving a general overview of the main mechanisms, with examples from suitable study systems. A great difference in length between these two chapters reflects the difference in research activity, with more studies being published on the benefits of grouping than on the costs. A discrepancy which we hope this book will help to remedy by stimulating more people to work on the latter. Both benefits and costs vary as a function of group size, which gave rise to the development of the optimal group size concept, the focus point of Chapter 4. This is followed by a review in Chapter 5 of the existing evidence for differences in the costs and benefits of different spatial positions within groups. We will discuss to which extent some group positions are more suitable for foraging but also more prone to predation and how animals deal with trade-offs between such conflicting trends. In Chapter 6 we return to the issue of how animals choose groups on the basis of the associated costs and benefits, but this time we consider the composition of groups.

This is a rather underexplored area of research which merits more attention in particular from theoreticians. Chapter 7 on the evolution of grouping concludes the part of the book that deals primarily with the functional side of grouping behaviour. A brief description of artificial selection experiments and their success at changing the grouping tendency of animals over just a few generations is followed by a review of the evidence for changes in grouping behaviour in the field. Another source of information on the evolution of grouping is the comparative method, which can be used to test for correlations between ecological conditions and the social organization in groups of closely related species. In Chapter 8, we provide an overview over some of the most important environmental effects on grouping behaviour. We start by looking at the ontogenetic constraints on grouping behaviour, then continue with a number of case studies that illustrate the important role of neurotransmitters and hormones for short-term changes in sociality primarily among invertebrates. This is followed by a section on the importance of the different forms of learning for grouping behaviour, and we finish this chapter by reviewing some of the evidence for parasite-induced changes in shoaling tendencies in several species of fish. Chapter 9 on the mechanisms of grouping explores some ideas on how animals recognize suitable group mates and form groups in the first place. A detailed discussion of the mechanisms by which group members regulate their inter-individual distance leads into the topic of individual-based models. This is a particularly popular and powerful technique (often referred to in the context of self-organization) which helps us to understand how complex phenomena at the group level (such as nest building in ants or collective locomotion in bird flocks) can be understood on the basis of multiple often simultaneous interactions between individual group members. In Chapter 10, we try to bring all the major ideas of this book together to provide a guide to interesting and underexplored questions that will hopefully provide starting points for new enquiries.

2

The benefits of group formation

2.1 Introduction

In this chapter we focus on mechanisms that confer benefits to group members. We concentrate on mechanisms, rather than ecological circumstances. For example, we do not discuss the benefits of colonial breeding to sea-birds in a self-contained section. However, someone interested in coloniality in sea-birds should be able to identify the mechanisms that could potentially have a bearing on this. We hope that this mechanisms-based approach allows us to focus on fundamental issues rather than species-specific consequences.

We have tried to focus on generally applicable benefits and have thus omitted some with narrower taxonomic relevance. However some of these deserve brief mention. One particular area of great current research activity is that of breeding systems that cause adults other than the parents to contribute to the rearing of off-spring. Good overviews of this can be found in Hatchwell and Komdeur (2000), Cockburn (1998), and Emlen (1991). Other benefits of alloparental care are discussed by Riedman (1982). The most extreme case of such alloparental care occurs in social insects, for which we refer the reader to another book in this series by Crozier and Pamilo (1996), and to Chapter 7 of this book for evolutionary consider-ations related to alloparental care. We similarly will not consider the structures (such as bridges) that groups of insects can build from their own bodies; a starting point for discussion of this topic is Anderson and Franks (2001). One recently discovered example of such structures is that blister beetle larvae apparently aggregate to mimic the appearance of a female bee. This allows effective transmission on to male bees, which attempt to copulate with the mass. In this way, larvae are transported back to the bee's nest, where they develop (Hafernik and Saul-Gershenz 2000). Another field that we considered too specialist for this book is coalition formation in social prim-ates, for which a good starting point would be Packer (1977) or Dugatkin (1997).

This is by far the biggest chapter in the book, so some explanation of the structure is required (see Table 2.1). Sections 2.2–2.4 deal with different aspects of grouping to avoid predators, Section 2.5 is about finding food, Section 2.6 about finding a mate, Section 2.7 about keeping warm, and Section 2.8 about travelling more efficiently.

Section 2.2 is on the effect of grouping as a means of detecting approaching pred-ators. The idea that 'many-eyes' allow groups to spot predators more effectively is a bedrock of ecology textbooks, but we will argue that many of the assumptions of this theory lack strong empirical foundations, and the benefits of grouping via

Table 2.1 Summary of the potential benefits of grouping that are considered in this chapter

Benefits	Study system	Authors
Anti-predator		
Many-eyes effect	Aquatic insects	Treherne and Foster (1980)
Encounter–dilution	Aquatic insect larvae	Wrona and Dixon (1991)
Predator confusion	Gulls	Fels *et al.* (1995)
Predator swamping	Mayfly	Sweeney and Vannote (1982)
Selfish herd effects	Theory	Hamilton (1971)
Defence against parasites	Horses	Rubenstein and Hohmann (1989)
Communal defence	Lions	Bertram (1975)
Predator learning	Insect larvae	Gagliardo and Guilford (1993)
Foraging		
Group hunting	African wild dogs	Creel and Creel (1995)
Coarse-level local enhancement	Geese	Drent and Swierstra (1977)
Fine-level local enhancement	Starlings	Templeton and Giraldeau (1995a,b)
Public information	Starlings	Templeton and Giraldeau (1996)
Information centre	Swallows	Brown (1986)
Mate choice		Westneat *et al.* (2000)
Keeping warm		
Reduced heat loss	Mice	Andrews and Belknap (1986)
Protection from desiccation	Butterfly eggs	Clark and Faeth (1998)
Reduced cost of transport		
In air	Pelicans	Weimerskirch *et al.* (2001)
On water	Ducklings	Fish (1991)
Under water	Fish	Herskin and Steffensen (1988)

reduced vigilance costs and/or improved early warning are more complicated than first appears.

Although the many-eyes hypothesis is the most commonly investigated aspect of the anti-predatory benefit of grouping, it is not the only one. Safety in numbers is another important and intuitively appealing concept. However, there are several different facets to the basic idea that an individual is less likely to be the one targeted by a predator in a big group than in a smaller one. These are discussed in Section 2.3. First there is simple dilution; if there are N individuals in a group and the predator picks at random, then your chance of being targeted is $1/N$. Then there is encounter–dilution, the idea that predators might encounter prey less frequently if the prey are bunched. However, these two effects are closely interrelated and it is much more effective to think of their joint action, so called attack-abatement. Predators will generally have some upper limit to their prey capture rate. In such circumstances, it may benefit prey to co-ordinate their activities so as to be exposed to the predator at the same time, in order to take advantage of this limitation in capture rate. Another commonly used phrase is that of the 'Selfish herd'; this phrase is often misapplied, and is presented in Section 2.3.3 to emphasize its difference from simple dilution. Lastly, we argue that many of the effects discussed in this section provide anti-parasitic defences as readily as anti-predatory ones.

In Section 2.4 we discuss the predator confusion effect. The simultaneous flight of a large number of individuals may make it difficult for a predator to focus on a particular target. Further, standing and fighting—rather than fleeing—may be an option for some groups of potential prey, where it would not be practical for an individual. Lastly, there has been repeated suggestion that aggregation makes predator learning of warning coloration more effective, although the mechanism by which this occurs remains unclear.

The benefits that can be obtained by group foraging are explored in Section 2.5. The idea that mobile predators can gain advantages from attacking in a group (Section 2.5.1) is firmly established in the literature. Slightly less obvious are the benefits that sit-and-wait predators like spiders can gain from coloniality, but this has been subject to significant research. This section also mentions the potential advantages of grouping to herbivores in order to overcome plant defences. The idea that foragers can gain information on where to find food from observation of others is a complex one. Particularly so, if we want to understand the spatial scales over which this mechanism operates and how individuals collect and process the information available. Hence Section 2.5.2 details different potential forms of information transfer and the empirical evidence for each.

Sexual selection is the current hot topic in Behavioural Ecology, and so it is no surprise that the benefits to mate choice of grouping have received a great deal of current interest, as reviewed in Section 2.6. For the choosy sex, grouping reduces the costs of sampling, for the more preferred individuals of the other sex, easier sampling is likely to benefit them also. The challenge is to explain why less-preferred individuals also participate in grouping for mate choice reasons.

There can be little doubt that huddling together can reduce heat or water loss. However, as shown in Section 2.7, we are still some way from being able to predict how these benefits will change with group size. Similarly, the idea (discussed in Section 2.8) that the costs of transport can be reduced by grouping is well known from human sports. Surprisingly, although schooling fish seem at first sight to be a classical example of this, clear empirical evidence of an energy saving in this case is lacking, and agreement between theory and observation on the expected relative positioning of fish is poor.

2.2 Anti-predator vigilance

2.2.1 The classical many-eyes theory

Several studies have demonstrated that larger groups are more effective at detecting approaching predators (Powell 1974; Siegfried and Underhill 1975; Kenwood 1978; Lazarus 1979; Treherne and Foster 1980; van Schaik et al. 1983; Godin et al. 1988; Cresswell 1994; but see Godin and Morgan 1985). These studies generally scored effectiveness as the frequency with which a predator-like stimulus elicited a visible response in prey or the distance at which an approaching threat did so (Fig. 2.1).

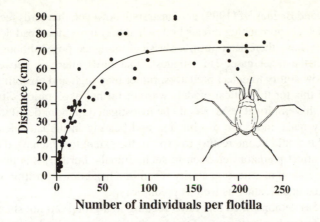

Fig. 2.1 The relationship between the number of ocean skaters in a flotilla and the distance at which responses to an approaching model predator were first observed. (Adapted from Treherne and Foster 1980.)

Both of these measures suffer from a methodological problem that there may be a delay between detection and visible response (Godin *et al*. 1988; Ydenberg and Dill 1986). However, fortunately it seems likely that such a delay will tend to increase with group size, if individuals in larger groups consider themselves safer. If delays in responding after detection were important, then this would lead to a tendency for response times to increase with group size; the opposite of what is observed. Hence, it does seem likely that individuals in bigger groups do actually detect predators earlier in an attack, compared with those in smaller groups.

Leaving this methodological problem aside, the group size effect is generally considered to be a consequence of an increased number of individuals scanning for predators, coupled with transmission of information about a detected threat throughout the group. Hence an individual in a group need not detect a predator itself to be warned of an attack, so long as at least one of the other group members does and informs the others. This 'many-eyes' theory suggests that as group size increases, individuals can decrease their personal commitment to vigilance without increasing their risk of failing to detect an attack. Such a decrease in vigilance has been well documented empirically (see Roberts 1996 for a review), and widely predicted by theoretical models (Pulliam 1973; Pulliam *et al*. 1982; Parker and Hammerstein 1985).

2.2.2 How individual vigilance works

Recently, several of the underlying assumptions of the many-eyes theory have been challenged. Theoretical models generally assume that foragers alternate between periods of feeding (when a predator's approach cannot be detected) and scans (when it can). Lima and Bednekoff (1999) demonstrated, however, that birds feeding on the ground can detect approaching threats both when they have their head down feeding and when they have their head up in a classical vigilance pose, although detection was easier in the latter case. This suggests that birds alternate between periods of high and low ability to detect predators, rather than high and no ability. The consequences of this for theoretical models warrant further exploration. Similarly, the generality of this empirical result should be investigated in more detail. Mammalian prey probably make more use of olfaction and hearing in predator detection than birds do, and it would be interesting to explore the extent to which head down feeding postures affect predator detection in such animals. Further, it is likely that the extent of attenuation of detection ability when in a head down position will depend on local habitat and weather as well as predator type.

Lima (1995a) demonstrated that birds feeding in a group did not show any reaction to the differing vigilance rates of other group members. This lack of behavioural monitoring begs the question of why group members do not cheat by giving up their own vigilance and relying on the vigilance of others. The likely answer to this is that the vigilance of others is not entirely as good as being vigilant yourself. There are at least two potential reasons for this. First, predators may preferentially attack low-vigilance individuals within a group, as found by FitzGibbon (1989) for cheetah attacking Thomson's gazelle, and Krause and Godin (1996a) for cichlid fish predators attacking smaller fish prey (guppies). Secondly, vigilance may allow quicker reaction to the signal of another group member. Lima (1994a,b) designed experiments such that only one ground feeding bird in a group detected an approaching threat. The other individuals in the group were classified as to whether they had their head up or down at the moment that the informed bird flushed for cover. The 'head up' type responded significantly more quickly (by flushing themselves) than the others. The models of Packer and Abrams (1990) and McNamara and Houston (1992) demonstrate that such an advantage to personal vigilance allows for the formation of an evolutionarily stable non-zero level of vigilance within a group.

2.2.3 Information transfer between individuals

Another key assumption of the many-eyes theory is that information about impending attack is transmitted unambiguously between those group members that detect it and those that do not. However, Lima (1995b) demonstrated that sparrows were unable to distinguish between predator-driven departures by other group members and those induced by other factors. Rather, birds reacted to the position of the flying bird (being more likely to respond to a bird from the outer portion of the flock) and to the number of departures and the time interval between them. This may be a special feature of taking flight in birds, which needs considerable physical effort in most circumstances. Locomotion in mammals can be considerably more graded, so ambling into shade is likely to be easily differentiated from bolting away from a predator. Even in Lima's study where fright and non-fright responses could not be distinguished, there was evidence that some information was conveyed by the

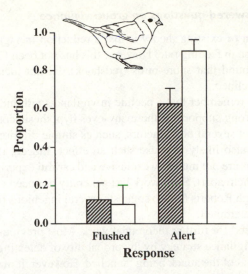

Fig. 2.2   The mean (±S.E.) proportion of non-detectors that flushed into cover or became alert following departures of the target bird induced by seeing either a ball (hatched bar) or a model raptor (open bar) approaching. (Adapted from Lima 1995b.)

fright-induced flight of another group member. After a target bird had been flushed, by either an oncoming ball or model raptor, more of the other birds immediately adopted an 'alert' pose in the latter case (Fig. 2.2). Hence, fright-induced departures and departures driven by other reasons did not elicit different proportions of departures in the test birds but did affect the proportion of alert postures. Davis (1975) demonstrated that the alarmed flight of a pigeon induced flock-mates to take flight in a way that non-alarmed departures did not. This was linked to a lack of characteristic pre-flight movements in the alarmed birds. Further, several studies using fish demonstrated that individuals respond to the 'alarmed' behaviour of others (Mathis *et al.* 1996; Ryer and Olla 1991; Godin *et al.* 1988; Krause 1993a; Magurran and Higham 1988). This last paper featured the ingenious use of mirrors such that the test fish could see the frightened conspecific but not the model predator. Similarly, Treherne and Foster (1981) observed that the approach of a model predator caused anti-predatory movements to spread across flotillas of a marine insect. Such information transfer is suggested to allow individuals to perform anti-predatory behaviour before the predator itself comes into their detection range. Treherne and Foster called this the Trafalgar Effect, after the signals transmitted by chains of ships at the Battle of Trafalgar allowing Admiral Nelson knowledge of events beyond his horizon. Further empirical work on the effectiveness of information transfer is needed, as well as consideration of existing empirical results for the theory (Lima 1994b).

2.2.4 Some unanswered questions on group vigilance

Another assumption of existing theory is that a reduction in vigilance allows for a concomitant increase in feeding rate. However this has not been fully tested. Indeed, Cresswell (1994) found that shore-birds (redshank) did not increase their feeding rate as vigilance declined.

It is important to remember that a decline in vigilance with increasing group size does not provide strong support for the many-eyes hypothesis considered here. The reason for this is that several other factors, such as simple dilution of risk discussed in Section 2.3, are also likely to cause such an effect. Indeed, the many-eyes and dilution mechanisms are not mutually exclusive and careful experimental design will be required to tell them apart. Such work has generally not been carried out (but see Dehn 1990), although Roberts (1996) suggests several ingenious pathways by which progress could be attempted.

There is also a need for further theoretical work. Most previous studies judge the success of a given vigilance strategy by the probability of detecting a predator within a certain time interval of the attack being launched. However, it may be that there are advantages to detecting the attack as early as possible, not considered by this theory. Similarly, there is a need to consider the decision-making process of the predator regarding the choice of which group to attack and which members of a given group (Bednekoff and Lima 1998). Further, flight is not the only way in which animals respond to information concerning a potential threat; alternatives such as immediate initiation of a scan (Lima 1995b) or increased vigilance rate (Roberts 1995), need also to be explored.

Another advantage to the better detection afforded by groups is that predators may sometimes break off attacks after prey have signalled that the predator has been detected. FitzGibbon (1989) reported that 52 out of 70 cheetah stalks were abandoned after the group of gazelle had indicated detection. Under such circumstances, the mechanisms that keep detection signals honest deserve some consideration. Recently, this question has been approached theoretically by Bergstrom and Lachmann (2001). They demonstrated that signalling can honestly reveal prey awareness of a predator provided the prey's assessment of predation risk accurately reflects the probability that an active predator actually is present and that greater awareness of the predator allows the prey a greater chance of escape from an attack. Both these conditions are likely to be met in many natural systems, and it may be that many alarm calls that have previously been thought of as signals to conspecifics act instead or additionally as signals of detection to predators. Bergstrom and Lachman provided several predictions that would allow empirical testing.

2.2.5 Related issues

A related problem to predator detection is the avoidance of fishing gears by fish. Brown and Warburton (1999) demonstrated that groups of five fish were more successful in escaping through a hole in an oncoming trawl than were groups of two.

This may be because more rapid detection of the trawl allows more time for fish to search for an escape route.

2.3 Dilution of risk

2.3.1 Avoidance, dilution, and abatement

Following Pitcher and Parrish (1993), we distinguish between predator avoidance (sometimes called encounter–dilution), dilution, and abatement, although the three mechanisms are interconnected. Predator avoidance can occur when the perceptual range of the predator is low, relative to the movement speeds of predator and prey, such that predators must search their environment for prey. If the prey are clumped, then this may increase the time predators go between encounters. However, whether this leads to reduced predation is far from clear. If the prey clumps are neither more nor less detectable than a single individual, and if predators capture all the prey in a discovered clump, then the predator's time-averaged feeding rate (and so the predation risk to the prey) is unaffected by aggregation. Hence, avoidance on its own does not necessarily always bring benefits to the prey. However, the variance in feeding rate experienced by the predator will increase with increasing aggregation, and it is conceivable that this might induce the predator to switch to an alternative food type.

If there is also a dilution effect, such that the predator can only eat a fraction of the group of prey (say only one individual) and the rest of the group can escape, then there can be a benefit to grouping. However, this must be set against consideration of the detectability of groups (see Section 3.2). If a group of N individuals is N times more detectable than a single individual, then even if the predator can only capture at most one individual from a detected group, there will be no advantage to the prey in grouping. Hence, the effectiveness of grouping as a predator avoidance strategy is connected to the relative detectability of groups of different size. It is difficult to generalize about this, but it is easy to find circumstances where detectability may increase faster than linearly with group size, and equally easy to find situations where it may increase slower than linearly, although hard empirical evidence is very much lacking (see Section 3.2).

We now look more closely at dilution. If a predator has discovered a group of prey, but can capture only one of these, then the larger the group the lower the chance that one particular individual will be the one attacked. Indeed, the probability might be expected to decline inversely with group size. There is evidence for this from Foster and Treherne's (1981) experiments with fish foraging on insects on the water surface, and Godin (1986), who found that the number of attacks by a predatory fish on a group of prey was independent of group size. In both cases, the authors interpreted their results as evidence for per capita predation risk to decline as $1/N$, where N is the group size. Similar results were obtained by Morgan and Colgan (1987) using approaches of a predatory fish to different sizes of prey shoals behind glass.

However, this interpretation requires several caveats. The 1/N rule assumes that each individual is equally likely to be the one attacked, a circumstance that will often not hold if predators preferentially target some types of individuals (see Chapter 6). Furthermore, there is some evidence that larger groups are more conspicuous to predators resulting in higher attack rates (see Chapter 3).

Further, not all experiments have produced support for the 1/N rule. Watt *et al.* (1997) presented toad tadpoles on a floating clear platform to a predatory fish. They found that strike rate did increase with group size, but sub-linearly, so that individuals in larger groups received proportionately fewer strikes. Fels *et al.* (1995) presented groups of croutons to circling gulls and found that the 'survival' probability of an individual crouton increased with group size, although not as quickly as the 1/N rule would predict. This was because gulls were in general more likely to capture an individual from a larger group (Fig. 2.3).

If a group of N individuals is N times more likely to be detected as a singleton (so there is no abatement effect), then dilution will be insufficient to overcome this disadvantage to grouping. Again we see that avoidance and dilution must be considered together. Ideas on avoidance and dilution received theoretical synthesis in the work of Turner and Pitcher (1986) which makes clear that avoidance and dilution should be studied in combination, as what they call the attack abatement effect (see Box 2.1). They argue that in circumstances where there is no advantage to group-living from either dilution or avoidance alone, an advantage can accrue from the two mechanisms working together. Wrona and Dixon (1991) discussed statistical methods for separating and comparing the effects of encounter reduction and dilution. They applied these to data on worm predation on the pupae of a stream-dwelling insect. They found that larger groups were more likely to be attacked than smaller groups

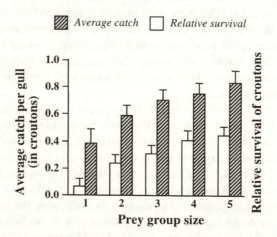

Fig. 2.3 The mean (±S.E.) number of croutons caught by a gull and the relative survival probability of croutons as functions of group size. (Adapted from Fels *et al.* 1995.)

(the opposite of the avoidance argument) but that within an attacked group, the probability that a focal individual was attacked was less in larger groups (revealing a dilution effect). They suggested that the second effect dominates the first, and so there is an overall attack abatement effect. Inman and Krebs (1987) emphasized that if there is some other advantage to being in a group, through an entirely separate mechanism perhaps even unrelated to protection from predation, then either dilution or avoidance acting alone can benefit group members. Hence, neither dilution or avoidance alone is enough to lead to the evolution of grouping (as discussed by Turner and Pitcher 1986), but can contribute to its maintenance once grouping has evolved by other means.

Arnqvist and Byström (1991) considered a model where predators are limited in some way that sets a ceiling on the number of individuals that can be captured when a group is discovered. They argued that this effect should lead to bimodal distributions of prey group sizes, since individuals within intermediate sized groups are less well protected than those in small or large groups. They find empirical support for this using parasitic wasps preying on clusters of water strider eggs.

Pitcher and Parish (1993) discussed another potential mechanism of dilution. The 'cognitive dilution effect' implies that when the predator can only attack one group member, the other group members gain information about the position, state, or capability of the predator that may be useful to them in future. As they pointed out, such an effect would be difficult to investigate experimentally.

Box 2.1   Attack abatement

Turner and Pitcher (1986) suggested that the risk of detection by a predator and the risk of being attacked need to be seen in context. The diagram on the next page shows the probability of detection during the search phase of the predator and the probability of being selected as a target by the predator during the attack phase. The combination of the two probabilities shows how the overall probability of being captured by a predator can be reduced by being in a group. Note however, that this model requires a number of assumptions. First, the probability of detection should not increase proportionately with group size (i.e. a group three times larger should not be detected three times more often). Otherwise the advantage gained through risk dilution is lost due to the higher detection probability. Secondly, given an attack, the predator's probability of successful capture of prey should be unaffected by prey group size. This point makes empirical studies of the encounter–dilution effect very difficult because animal groups often gain risk reduction through several anti-predator effects (e.g. predator confusion effect, many-eyes effect) simultaneously. However, in order to demonstrate the existence of the encounter–dilution effect we need to isolate it from other effects. Thirdly, the predator should only take a single prey given an attack. If the predator can take several individuals per attack, then grouping (on the basis of the encounter–dilution effect) would only reduce the predation risk for groups larger than the maximum number that the predator can eat on a given attack. This makes it difficult to explain the evolution of grouping because only individuals in large groups but not ones in small groups would derive a benefit.

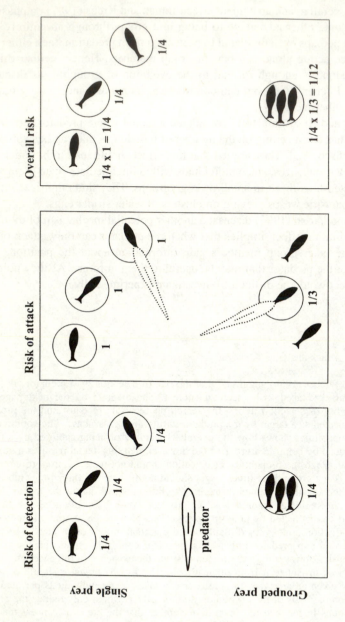

Figure for Box 2.1.

2.3.2 Predator swamping

It is likely that the rate at which a predator can catch prey will have some upper limit (e.g. because subduing a prey individual takes a finite time). Prey can take advantage of this constraint by synchronizing emergence which reduces the risk to the individual compared with when they present themselves individually over a longer time period.

This is thought to explain the observation of clustered emergence of bats from roosts. It has also been suggested as an explanation for synchrony of emergence of adult insects such as mayflies (Sweeney and Vannote 1982) and cicadas (Lloyd and Dybas 1966). Sweeney and Vannote give empirical evidence that the percentage of adult mayflies caught by predators is inversely related to the total number of adults available that day. An alternative explanation for synchrony could be an enhanced ability to find a mate, but these authors found that synchrony among parthenogenetic mayflies was equal to (or perhaps greater than) that of sexual species. Similarly both frogs and air-breathing fish have been demonstrated to increase the synchrony of their breathing following a disturbance (Gee 1980; Baird 1983).

2.3.3 The Selfish herd

Another related idea is that of the Selfish herd, attributed to Hamilton (1971) who envisaged a predator that captures the first prey item that comes within range. One can argue that an individual on its own is more vulnerable to such a predator than one in a group, because it has a greater 'domain of danger' (see also Section 5.2). This term means that there are more predator search trajectories that cause the lone individual to be discovered first, rather than an individual in a group, which gains protection from the vulnerability of those around it. This could be a strong selection pressure, which will lead to higher mortality of singletons than grouped individuals. However, exactly how such a mechanism influences the evolution of grouping is less clear. Morton *et al.* (1994) called into question that a simple rule biasing movement towards an individual's nearest neighbour was sufficient to explain the evolution of grouping under such an anti-predatory advantage of grouping, and further work on this would be valuable.

2.3.4 Defence against parasites

All of the mechanisms discussed in the preceding sections are equally as valid for defence against highly mobile parasites as against predators. Indeed, Mooring and Hart (1992) reviewed considerable empirical evidence both for abatement and Selfish herd effects against parasites. Grouping may, however, also incur a cost from less mobile contact-spread parasites (see Section 3.4). This receives support from work by Rubenstein and Hohmann (1989) who found that the numbers of biting flies on an individual decreased with herd size in feral horses, but infection with endoparasites increased. Poulin and FitzGerald (1989) conducted laboratory experiments using a free-swimming blood-sucking crustacean ectoparasite, preying on stickleback fish. They found that the frequency with which the ectoparasite attacked increased with the density of fish in the tank, but the success of these attacks was unaffected by density. Overall, the increase in attack rate was less than linear and so each individual fish was less at risk when in a larger group, consistent with an abatement effect. A further benefit was obtained from an increased chance of the parasite being eaten (and so destroyed) by one of the fish in larger shoals. Fish responded to parasite presence by an increase in shoaling tendency. Poulin (1999) followed this up with a field study in which he predicted that infection levels with this parasite should decrease with shoal size. Such a relationship was not found, perhaps due to the small sample size (14 shoals) or because the effect was confounded by other factors. Côté and Gross (1993) showed that fungal infection decreased with increasing density in bluegill sunfish nests (Fig. 2.4). This can partially be explained by males in larger colonies having to spend less time on territorial defence and thus being able to spend more time fanning the nests. However, experiments excluding male behaviour also showed this effect. The authors suggested that this may be an abatement effect, although a more plausible explanation may be a swamping effect, occurring if fungal spores can become locally depleted because of very slow transport through water currents in the lake. Rutberg (1987) found that feral ponies in larger groups suffered from fewer flying insects, and that spacing decreased (although group sizes remained unchanged) when flies were more common. Duncan and Vigne (1979) found a similar effect of group size on parasite infection rates in feral horses. Espmark and Langvatn (1979) reported that reindeer tend to reduce inter-individual distances when attacked by mosquitoes, but then increase these distances, when attacked by tabanid and oestrid flies. This may suggest that the different types

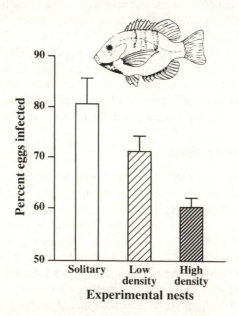

Fig. 2.4 The mean (±S.E.) proportion of bluegill eggs infected by fungus in experimental solitary nests and low and high density colonies. (Adapted from Côté and Gross 1993.)

of insect use different sensory means of locating potential prey. Côté and Poulin (1995) carried out a meta-analysis revealing that the intensity of infection with mobile parasites consistently declined with group size, and discuss many incidents when hosts respond to increased mobile parasite abundance by forming bigger groups. It may be significant, however, that all their mobile parasites share the feature of being satiated after exploiting one host.

2.4 Predator confusion

2.4.1 Theory

The confusion effect describes the reduced attack-to-kill ratio experienced by a predator resulting from an inability to single out and attack individual prey in a group. This idea received theoretical support from the artificial neural network model of Krakauer (1995). His abstract model of the computational mechanisms that may plausibly underlie perception of prey positions made the following predictions:

(a) All group members benefit from this protection.
(b) There is an exponential decrease in predator success with increasing group size (see Fig. 2.5).
(c) Increased protection occurs following group compaction.
(d) The confusion effect is most effective when all individuals are alike, with odd individuals suffering disproportionately from predation.

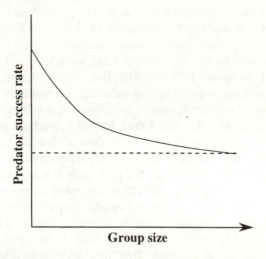

Fig. 2.5 Krakauer's (1995) model predicts that the predator success rate should decline with increasing group size, but in a decelerating way, with each additional individual added to the group having less effect than the last.

2.4.2 Empirical support for theoretical predictions

We defer further consideration of the effects of homogeneity and oddity to Chapter 6. Other empirical support for the confusion effect is limited, perhaps because the design of experiments that rule out other confounding factors is challenging. Milinski (1977a,b) found that sticklebacks preferred to attack straying water fleas to aggregations. Shoaling fish have been observed to show a compaction response following an increase in perceived predation risk (Seghers 1974; Magurran and Pitcher 1987). Further, Milinski (1990) demonstrated an increase in the time needed to pinpoint a fixed number of targets on a screen by humans when the density (but not the number) of targets went up.

There are also numerous suggestions that predators attempt to disrupt aggregations in order to isolate potential targets (Major 1978; Schmidt and Strand 1982; Schaller 1972), however, the extent to which predators benefit from such isolation remains unclear. Page and Whitacre (1975) reported that birds of prey (merlin) were more successful against some group sizes of small shore-birds (sandpipers) than others. However such results are difficult to interpret, since group size is likely to be influenced by factors (such as weather) that are also likely to affect predation risk. Kenwood (1978) also showed a strong decrease in the percentage of hawk attacks that were successful as pigeon flock size increased. However, he did not attribute this to the confusion effect, but rather to weaker birds tending to forage in smaller groups (away from competition) and smaller groups being less able to detect oncoming attacks.

Many of these confounding factors are controlled for by the laboratory experiments of Neill and Cullen (1974), which demonstrated that four aquatic predators (squid, cuttlefish, pike, and perch) all had a reduced success rate per attack when attacking prey fish in groups, rather than singly (Fig. 2.6). Similarly, Treherne and Foster (1982) demonstrated that the probability of successful attacks by a fish predator decreased as group size of marine insects increased. Gillett *et al.* (1979) found that both humans and a lizard took longer to catch a single cricket in an arena as the number of crickets increased. Fels *et al.* (1995) directly tested the confusion effect, neatly avoiding many confounding factors and ethical problems by using croutons as prey. Groups of croutons were thrown in the air, and attacked by waiting gulls. No evidence for a confusion effect was found, indeed gulls did better against larger groups. One drawback to these experiments is their artificiality. Gulls and croutons are not a naturally evolved predator–prey system, although the authors suggest that members of the public have commonly fed gulls at their study site in a similar way to that used in the study. Schradin (2000) also eliminated many confounding factors by offering very simple prey (mealworms) to both leopard geckos and marmosets. However, whether this represents a foraging situation close to anything that these species would experience in nature is open to debate. Putting this issue to one side, both predators took longer to catch their first prey item from larger groups. Krakauer's most novel prediction, the decelerating drop to an asymptotic level of success with increasing group size, remains untested.

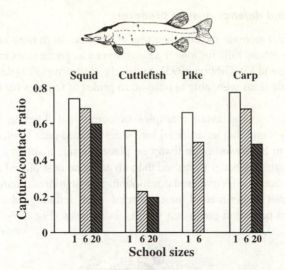

Fig. 2.6 The success rate per attack of squid, cuttlefish, pike, and perch when attacking prey fish in groups or singly. (Adapted from Neill and Cullen 1974.)

2.4.3 Cognitive limitations

Milinski and Heller (1978) demonstrated that hungry fish (sticklebacks) preferred to forage on large aggregations of water fleas, and achieved higher foraging rates feeding on larger aggregations than smaller ones. Conversely, well-fed fish preferred smaller aggregations and achieved higher uptake rates on these than on larger aggregations. They suggested that these results occur because predator confusion effects mean that fish must devote a greater proportion of their attention to foraging when dealing with larger aggregations, and this can only be achieved by reducing anti-predatory vigilance. Hungry fish were prepared to pay this price, but well-fed ones were not. In support of this, they found that flying a model kingfisher (a potential avian predator of sticklebacks) over the tank led to a shift in predation towards smaller aggregations by hungry fish. Further, they found that fish feeding on small prey aggregations reacted to both conspicuous and cryptic model predators. The same was true for the conspicuous model predator among fish feeding on large aggregations, but the fastest feeding of these fish failed to react to the cryptic model predators. Ohguchi (1981) demonstrated that sticklebacks would attack water fleas much more readily than similar sized and coloured carrot pieces, when they were presented singly. However, when one of each were presented simultaneously, both were attacked with equal frequency. This experiment suggests that the ability to recognize prey is compromised when multiple prey objects are present.

2.4.4 Communal defence against predators

Grouping can also increase the effectiveness of defence. Both lions and African wild dogs are able to defend kills for longer against hyena as group sizes increase (Cooper 1991; Fanshawe and FitzGibbon 1993). Similarly, Bertram (1975) demonstrated that coalitions of male lions were able to hold on to prides of females for longer than single males.

There are also several avian examples of communal defence. Andersson and Wiklund (1978) found that an artificial bird's nest placed near a colony of fieldfares was less likely to be preyed upon than one placed beside a solitary active fieldfare nest, which in turn was better protected than an artificial nest placed alone. This was suggested to be due to effective predator mobbing by fieldfares. Picman *et al.* (1988) demonstrated that experimental nests placed near redwing blackbird colonies suffered reduced predation compared with isolated nests (Fig. 2.7). This again was

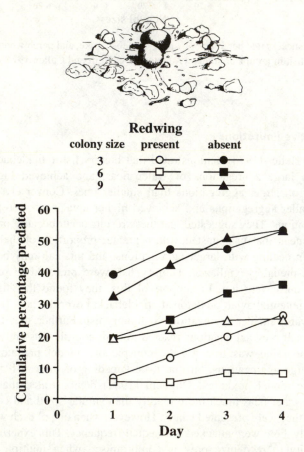

Fig. 2.7 The cumulative percentage of artificial nests predated on days 1–4 after baiting, as functions of the colony size in which artificial nests were placed, and whether or not the colony was placed around an active redwing nest. (Adapted from Picman *et al.* 1988.)

attributed to the protection afforded by mobbing. Similarly mobbing was shown to be more effective in larger colonies of yellow-rumped caciques by Robinson (1985), who also referenced several other studies showing similar effects. Hoogland and Sherman (1976) reported mobbing by larger numbers of swallows nearer larger colonies. A similar effect has been shown for mobbing as defence against parasitism by cowbirds (Clotfelter and Yasukawa 1999, and references therein).

Mobbing behaviour can certainly be dangerous for the mobbers (Denson 1979; Poiani and Yorke 1989), so there must be fitness benefits sufficient to compensate for these costs, both for the initiator of mobbing and for those that join. Curio (1978) listed several potential benefits. In some cases, the potential prey can succeed in killing the would-be predator (Furrer 1975). Another function may be to simply communicate to a predator that it has been discovered, which may be sufficient to cause the predator to abandon its attack. More persistent harrying may make such abandonment more likely, and Andersson (1976) demonstrated that larger numbers of seabirds were more successful in driving predatory birds (skuas) away. There may be longer-term benefits in reducing the predator's likelihood of returning to the same place (Pettifor 1990; Flasskamp 1994), and/or depriving it of the ability to gather information that would improve the success of any further attacks. Alternatively or additionally, mobbing may serve to warn kin of approaching danger, or may serve to pass on information to naïve kin about dangerous species or locations. Several publications have presented evidence for some of these mechanisms (Curio *et al.* 1978; Frankenberg 1981) but critical testing between mechanisms is difficult, and may be fruitless, as multiple benefits are likely in many situations.

2.4.5 Predator learning

There is experimental evidence suggesting that the effectiveness of aposematic coloration as an anti-predator defence is enhanced by grouping (Gagliardo and Guilford 1993; Gamberale and Tullberg 1996, 1998). However, the mechanism by which aposematic defence is enhanced by grouping has yet to be fully understood, and the benefits of grouping in this regard have yet to be quantified. Such experiments will require a very careful design, as discussed by Lindström (1999).

2.5 Foraging benefits to grouping

2.5.1 Benefits for predators

Grouping can allow predators to capture prey types that would be too large, too agile, or dangerous for a single individual (Creel and Creel 1995; Fanshawe and FitzGibbon 1993; Gese *et al.* 1988; Hector 1986; Bednarz 1988). Protective groupings may be broken up by the simultaneous assault of several predators, or prey fleeing from one individual may become more obvious to others or may flee into their path.

Group hunting can involve complex behaviour where individuals adopt different mutually complementary roles and exercise temporary restraint of feeding behaviour until prey have been rendered more vulnerable. Such behaviour has been observed in yellowtail fish hunting various shoaling fish. They collect shoals and drive them into the shallows before attacking (Schmitt and Strand 1982). In other cases, such complex co-operation is not required. Duffy (1983) described the foraging of diving birds feeding on shoals of fish. He speculates that multiple predators disrupt the shoal and destroy the fish-to-fish co-ordination required for synchronous evasive actions, and may cause fish to become exhausted through overuse of anaerobic muscles. Götmark *et al.* (1986) demonstrated that the fishing success of black-headed gulls diving for such shoaling fish increases as group size increased; similar results were obtained for predatory fish by Major (1978) (see Fig. 2.8).

Hunting in groups can also provide access to defended food sources. Flocks of jackdaws (members of the crow family) can secure food items from much larger crows that would dominate a single bird (Röell 1978). Similarly, flocks of juvenile ravens (also members of the crow family) can overcome the defence of a territorial mated pair of adults and gain access to a carcass (Heinrich and Marzluff 1995). Similar effects have been observed in shoaling fish (Foster 1985).

Some spiders live colonially, building separate nests in very close proximity to each other. One benefit to this may be that between-web transmission of vibrations

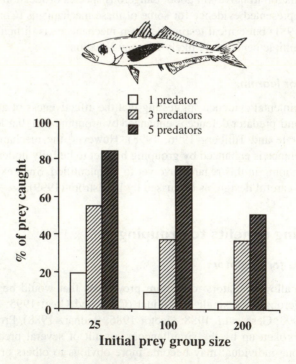

Fig. 2.8 The percentage of anchovies caught by predatory jack as a function of the numbers of each in an experimental pen. (Adapted from Major 1978.)

caused by struggling prey may allow individuals to track spatial variation in prey availability, and so make better informed judgements on relocation. Alternatively, or additionally, coloniality may increase the number and/or size of prey items caught in each individual web through the 'ricochet effect', where a prey item may escape from one web, but be more vulnerable to subsequent capture in another web through being disorientated or partially entangled in web fragments. Further, food stealing from the webs of others may allow individuals to reduce the variance in uptake. This may explain why such colonies dissolve when prey is scarce. Theory would predict that individuals should become more risk prone in such circumstances, choosing to forage alone, increasing the variance in food supply. In some species, spiders band together in the construction of a single web, sharing prey captures, thus potentially saving on construction costs compared with the separate webs of colonials. An introduction to this literature can be found in Caraco *et al.* (1995).

Le Masurier (1994) argued that herbivorous insects can exploit food better by being aggregated and targeting host plants sequentially. The mechanism behind this is that plants respond to attack by reducing growth. By attacking plants sequentially and not simultaneously, insects can take advantage of the high initial growth rate of plants and thereby increase net-energy intake. Additionally, aggregations of herbivorous insects can overwhelm induced chemical defences in host plants (e.g. Denno and Benrey 1997).

2.5.2 Finding food

Although we focus on food finding here, similar arguments apply to other resources, such as mates (Nordell and Valone 1998).

2.5.2.1 Joining behaviour or coarse-level local enhancement

The behaviour of other group members is a potential source of information to an individual searching for food. Here we summarize what is known about the extent to which such information is used, and how useful it is.

It is well documented that a forager or group of foragers can attract others to the site of their foraging. Such situations are likely to occur when food is distributed in ephemeral patches, which can be shared by more than one individual. This situation is well known to humans, the easiest way to find dead ungulates on the African plains is to home in on the plumes of vultures that circle above them. There is evidence, at least in some taxa, that foraging conspecifics are more attractive than those engaged in other activities. Krebs (1974) investigated whether herons were attracted to feed near model herons. He mentioned that preliminary tests indicated that a model in a hunched neck roosting posture was less attractive than an upright model. This may indicate that birds are preferentially attracted to apparently feeding individuals, or it may simply be that such individuals are more visible. More convincing are the results of Drent and Swierstra (1977). They compared two different types of model geese flocks, one had 15 models in a head down grazing posture and 15 in a head up alert posture; the other had 25 grazers and five alert individuals. This second group attracted over twice as many geese flocks to land near it. Hence, it is clear that geese used information not just on the presence of conspecifics, but also on their apparent behaviour when deciding whether or not to land. Similar results were obtained by Inglis and Isaacson (1978) using geese, and Murton *et al.* (1974) using woodpigeons. In contrast, the posture of real teal did not affect the attractiveness of groups in the study of Pöysä (1991).

Drent and Swierstra (1977) also explored the influence of group size on attractiveness, but found no effect. However they cautioned against reading too much into this, as they suspected that their experiment was confounded by the use of both alert and feeding models. One also has to be careful in interpreting natural situations where individuals choose to join the larger of two foraging groups because this need not be evidence of cueing on group size. It may be that individuals select the better food site, which then tends to have the larger group of foragers on it. Reebs and Gallant (1997) reported that hungry fish preferred to shoal with conspecifics showing food anticipatory behaviour. Similarly, Lachlan *et al.* (1998) showed that guppies preferred to shoal with conspecifics with higher local foraging experience. This may be related to their ability to find food through social learning (Laland and Williams 1997).

Theoretical models predict that joining a group will either have no effect or a negative effect on an individual's net mean foraging rate, but its real advantage occurs in reducing the variance in times between obtaining food (Ruxton *et al.* 1995). It is no surprise that net mean foraging rate does not increase, as such joining behaviour does not increase the ability of a set of individuals to discover new food patches (compared with the aggregate of the same number of individuals searching independently). Net mean intake rate can go down because of interference at foraging sites, or because new food patches are found less quickly than would be predicted for an equivalent number of independent searchers, because aggregation of individuals tends to lead to overlap of their search areas.

In heterogeneous groups, it may be that dominant individuals, that can obtain more than their fair share of finds, can exploit such a system to boost their mean reward rate, but when individuals are effectively identical, then mean reward rate will probably decrease. This is not to say that such behaviour would be selectively disadvantageous, as an individual that did not join in a population of joiners would often do worse, whereas an individual that exploited the discoveries of others in a population of non-exploiters would often do better. The costs and benefits of such behaviours are frequency-dependent and so a game-theoretical approach would be appropriate and potentially illuminating. Such exploitation behaviour in individuals that do not form a close-knit group has been termed coarse-level local enhancement by Poysa (1992).

2.5.2.2 Fine-level local enhancement

Poysa (1992) described fine-level local enhancement as the exploitation of the discoveries of other group members within a cohesive group that always stays together. Conceptually one might differentiate this from coarse-level local enhancement by considering it to apply to situations where the distances between individuals are sufficiently small that the relative positioning of individuals within the group can be ignored. Such assumptions have lead to extensive development of producer–scrounger theory (Giraldeau and Beauchamp 1999; Giraldeau and Caraco 2000). Giraldeau and Beauchamp (1999) distinguished between producer–scrounger models, where producing and scrounging are mutually exclusive activities, and 'information sharing models' like Ruxton *et al.* (1995), where searching for food and for other foragers to exploit are entirely compatible. As with coarse-level local enhancement, definitive testing using observations of naturally occurring events can be difficult, because of uncertainty as to which cues individuals actually use for decision making.

Templeton and Giraldeau (1995a), however, presented results of a clever laboratory experiment. Birds sample feeding stations by breaking through a sequence of two seals; behind the first seal is a piece of coloured paper, behind the second seal is a compartment which may or may not contain food items (Fig. 2.9). A bird can

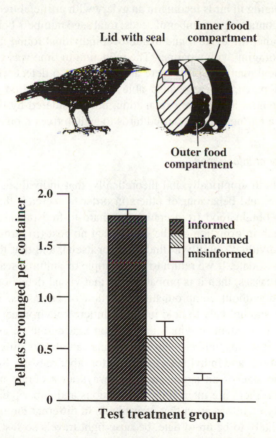

Fig. 2.9 The mean number of pellets scrounged by the test bird when the companion bird found a profitable container. (Adapted from Templeton and Giraldeau 1995a.)

improve its food gathering by sharing food discoveries of others. Templeton and Giraldeau compared the performance of three types of birds. One (the 'informed' type), was trained so that the colour of the tissue paper was a reliable indicator of whether or not there was food behind the second seal. The second type (the 'misinformed' one), was trained such that the colours used in the experiment misinformed it (the paper colour associated with food in training was associated with no food in the experiment and vice versa). The colours provided no information for the third type (the 'uninformed' one). Each foraged with an uninformed bird. It was found that the 'informed' type was able to take advantage of scrounging opportunities better than the 'uninformed' type which in turn did better than the 'misinformed' type (Fig. 2.9). Pitcher *et al.* (1982), demonstrated similar scrounging behaviour in minnows and goldfish.

Krebs *et al.* (1972) demonstrated using captive birds foraging in an aviary with different types of food sources that birds would use information from the food finding of others to bias search towards similar types of food sources. Benkman (1988) investigated foraging in birds feeding in an aviary with artificial trees, each of which had pine cones containing a number of seeds: total seed numbers being 49 in one tree and five in the other three. He found that a focal individual found its way to the best tree quicker if foraging with a partner. The reduction in time was greater than 50%, suggesting to Benkman that birds were able not only to detect when their partner found the food patch, but were also able to gain information on unsuccessful sampling. A potential alternative explanation, not considered by the author, is that the presence of a partner simply caused birds to switch trees more readily.

2.5.2.3 Reliability of information

It seems clear, both empirically and theoretically, that individuals use information from the position and behaviour of others in order to increase the frequency with which they can obtain food from ephemeral, hard-to-find patches. It is generally assumed that such information can be obtained at no cost (in terms of detriment to the observed individual's ability to find food for itself), and that the information is accurate and up-to-date. If we return to the example of vultures searching the plains for mammal carcasses, then it is probably true that visual detection of conspecifics can be achieved without compromising detection of carcasses. However, this is largely untested and unlikely to hold in more cluttered environments, especially for terrestrial taxa. The extent to which information is accurate is, again, empirically unexplored, but it seems unlikely that individuals can always differentiate between feeding aggregations and individuals gathering for other reasons. Similarly, it seems unlikely that the size of an aggregation is always an accurate measure of patch quality. However, there are interesting observations and experiments to be done to probe the quality of information that is available in different circumstances. Visual information is likely to be up-to-date, because light travels so fast, but this may be less true of detection by olfaction say. Consider deep sea scavengers. The opening of a carcass by some scavengers is likely to cause the gradual spread of suspended particles away from the site. These can be used by others to home in on the food source, but may remain as a piece of misinformation long after the food source has been entirely consumed. Even when a vulture detects a plume of circling conspecifics, it may still have to invest a significant amount of time in travelling to the site of the plume, which may be tens of kilometres away.

As well as empirical experiments, there is a real need for theoretical work to explore the consequences of less than perfect information. One key aspect to this may be the cost of making mistakes. Returning to the vulture one last time, a vulture might turn towards a plume of smoke in the far distance, mistaking it for a plume of conspecifics. As it draws closer, the mistake is realized. However the cost of this mistake may be very low, as the individual will have been able to continue searching for other foraging opportunities whilst in transit, and this search path may be just as likely to bear fruit as the one it would have adopted, had it not decided to divert.

2.5.2.4 Public information

We might next ask whether birds already on a patch can gain information on the quality of that patch from the behaviour of others. The situation we have in mind here, is a restricted area that contains an unknown number of cryptic food items. Individuals must decide when to leave a patch (perhaps because it is intrinsically poor or exhausted) and search for another. Valone (1989, 1993) developed a model to explore the benefits of such 'public information'. He argued that individuals in a group of foragers can gain a better estimate of the quality of a patch by using information about food discoveries by others as well as themselves. Empirical evidence for what foragers actually do with public information, coming entirely from avian studies, is limited but tantalizing.

Templeton and Giraldeau (1996) performed experiments on starlings feeding on artificial food patches in a laboratory. The experimental environment consisted of two patches each with two parallel rows of 15 holes. These holes were covered such that birds could visually inspect them for food, but had to manipulate the cover with their beak in order to sample (Fig. 2.10). The two patches were separated by a transparent barrier. Each of the six subject birds was trained to expect the 30 holes in a patch to contain either three or no food items between them, and that the two patches always contain the same number of food items (zero or three), but not in the same positions. Birds were tested in situations where there were no food items in the patches. Three situations were compared. In the first, the bird foraged in a patch alone with no bird in the other patch, and the number of empty holes investigated before the bird departs was recorded. The same was done in a situation where another bird was simultaneously feeding on the other patch. In contrast to the subject birds, this other bird had been trained to expect that food was available in three specified holes, so it probed a small number of holes before departing. This bird was termed the 'low information partner'. In another experimental treatment, subject birds were also allowed to feed simultaneously with a 'high information partner' in the other patch. The latter individual had been trained to expect only one of the holes

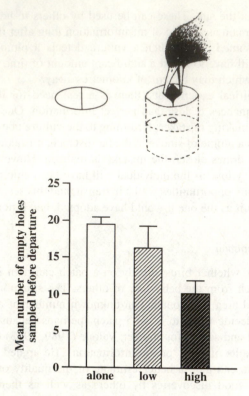

Fig. 2.10 The number of empty holes probed by the test bird when alone and when with low information and high information partners. (Adapted from Templeton and Giraldeau 1996.)

(at random) to provide food, so probed a large number of holes before departing. The experimenters' prediction was that, if the subject bird is using public information, then it should test most holes before giving up when it is alone, it should test fewer holes when it is with the 'low information' partner, and fewer still when it is with the 'high information' partner.

The initial experiment found no difference between the number of holes probed by the subject in the three situations. The experimenters then altered the shape of the patches, from a 2×15 array to a more compact 6×5 array. In this situation, the results were as they had predicted, and so consistent with the use of public information by subject birds. They interpreted the different results for different shaped food patches by suggesting that it was easier for the subject bird to keep the partner bird in sight in the 5×6 patch, without compromising its own foraging.

Smith *et al.* (1999) investigated the foraging of red crossbills in an experimental aviary. The aviary was divided into two equal halves. Two experimental feeding stations (termed trees) were also bisected by this central divider. Each feeding station presented several containers (cones), which birds had to inspect to find food. Birds were trained to expect some of the cones on one of the trees (at random) to yield food, and none of the cones on the other tree. They compared a single bird feeding in one-half of the aviary with a situation where two other individuals were able to feed in the other half of the aviary. Results were consistent with the use of public information. When feeding on the foodless tree, focal birds spent less time and sampled fewer cones before giving up when they were with two other foragers. The variances in these two measures also decreased. When the focal bird's side of the tree was empty, but the other birds could find food on their side, then focal birds searched more cones than when alone. In the mirror situation, where the focal bird's side had food but the other birds' side did not, then focal birds sampled fewer cones than when alone.

Templeton and Giraldeau (1995b) explored patch success by wild-living starlings trained to search for cheese pellets buried in sand within an array of plastic cups. The level of the sand was manipulated, so that in one treatment birds could always keep their eyes above the cup rim during foraging, but not in the other treatment. In the first treatment, foraging and gathering of public information were considered to be compatible, but not in the second. The authors looked at the order in which two individuals that started searching the array of cups simultaneously, departed. When there was no access to public information the less successful forager left first in 83% of the cases. In contrast, when the birds could view each other during foraging, then this was true in only 47% of the cases. These results were interpreted as evidence for the use of public information in the first case, such that each bird had the same estimate of the value of the patch at all times. Another related issue is the use of conspecifics as a cue in the selection of breeding sites. Brown *et al.* (2000) suggested that cliff sparrows probably use conspecific breeding performance in selecting colonies. Their evidence for this was a positive correlation between measures of per capita breeding success at a colony in one year and estimates of the net number of immigrants breeding there the next year. However, an alternative mechanism that could give rise to the same correlation is a situation where a bird's propensity to leave a colony decreases with increasing success, and after deciding to leave, another colony is picked entirely randomly. Unlike the authors' suggested explanation, this alternative involves no use of public information.

2.5.2.5 Information centres

The information centre theory was originally conceived by Ward and Zahavi (1973). This theory applies to situations where individuals that have found a food source return to a roost or colony before making a trip back to the food source. Assuming that this is the case, and that other colony members can identify successful foragers, it should be possible for them to follow such informed individuals to food locations. This theory has been subject to extensive empirical investigation (reviewed by Mock *et al.* 1988; Richner and Heeb 1995). These reviews highlight that many studies purporting to support the theory fail to consider other alternative explanations, particularly local enhancement. Richner and Heeb (1995) found only one field study (Brown 1986, on cliff swallows) that convincingly demonstrated an information centre mechanism. They also found laboratory evidence that information centre behaviour was possible in rats and honey bees. Hence it appears that the information centre mechanism does operate in some circumstances, but may not be common in nature. Although empiricists are responding to the challenge, and have produced experiments that avoid the methodological limitations of past work and provide evidence for information centres (e.g. the hooded crow roosts of Sonerud *et al.* 2001).

The view that information centres are relatively uncommon has recently been challenged on the basis of mathematical modelling by Barta and Giraldeau (2001). Their model predicts that avian breeding colonies can function as information centres. They further suggest that empirical tests of the information centre hypothesis should focus on different information from that traditionally gathered. They predict that instances where an uninformed individual follows an informed one will be relatively uncommon, hence failure to find such behaviour empirically is not good evidence against the information centre hypothesis. Richner and Heeb (1996) propose an alternative 'recruitment centre' mechanism, in which successful individuals return to the nest to recruit other individuals to a food patch because the benefits of being in a group outweigh the costs (including the costs of recruiting others). Functionally this is similar to food calling (see Chapter 4) but fundamentally different from the information centre hypothesis in that successful individuals benefit from recruiting others, rather than suffering competition from them.

2.6 Finding a mate

Leks are communal mating grounds in which males display and females visit to mate, gaining no resources other than sperm from attending the lek. They occur in species where males are unable to monopolize females or the resources needed for breeding. This has been the subject of intense theoretical and empirical interest: an introduction can be found in Chapter 7 of Sutherland (1996), and more comprehensive treatments in Wiley (1991) and Höglund and Alatalo (1995).

The attraction of lekking to females is likely to be that lekking allows choice between a large number of potential mates, and also allows effective assessment of individual male's quality. A simple sampling argument suggests that the best male of a large group is, on average, better than the best male of a small group (Janetos 1980; Kokko 1997). Sutherland (1996) also suggested that competition is likely to produce a correlation between the size of a lek and the *average* quality of males present. If males furthermore have to compete for the best sites on a lek, the position of individuals within a lek (with better quality males holding the central territories) can become a very effective measure of quality (Fiske *et al.* 1998; Kokko *et al.* 1999a,b). Although females may also prefer to mate on leks for other reasons, e.g. to reduce harassment by males (Clutton-Brock *et al.* 1992).

What is the benefit to males? It has been suggested that if females are more attracted to larger leks and more likely to copulate on each visit (for the reasons given above), then the average number of copulations per male will increase with lek size. This is, however, not observed for all lekking species (Sutherland 1996). Even where an increase is observed, these copulations are not evenly distributed, but are skewed, such that some males attract more than their fair share (Bradbury *et al.* 1985; Mackenzie *et al.* 1995; Kokko *et al.* 1998, 1999a). We must now explain why the unsuccessful males participate in leks under such conditions. Three suggestions have been proposed to solve this puzzle.

First, it may be that the increased total number of females copulating compensates for this skew, such that even less preferred individuals get more copulations by remaining in the lek than they would from dispersing (Kokko 1997). Widemo and Owens (1995) suggested that larger lek sizes reduce the extent of this skew, because increased male–male competition, increased disruption to copulation, and increased female arrival rates all serve to reduce the ability of the most preferred individuals to monopolize matings (Fig. 2.11). Errors in female choice can likewise reduce the skew (Johnstone and Earn 1999; Randerson *et al.* 2000). Hence it may be that the less preferred individuals actually benefit from being in larger aggregations. Despite the costs that a large aggregation imposes on preferred individuals, they too still do better in large aggregations than they would do on their own, and so remain in the aggregations. Tests of this argument are hampered by lack of suitable data, and remain inconclusive (Kokko *et al.* 1998).

Secondly, kin selection may play a role in lek evolution: if female copulation rate increases with lek size, an otherwise unsuccessful male can gain indirect fitness by attracting females to copulate with his brother, uncle, or father (Kokko and Lindström 1996). Data on several avian species appear to support the kin selection model (Höglund *et al.* 1999; Petrie *et al.* 1999; Shorey *et al.* 2000).

Thirdly, lekking may be an investment in future reproduction. If competition for the best positions involves an element of queuing, then a currently unsuccessful male may still benefit from lekking if it improves its access to better positions/spots within the lek in the future (Kokko *et* al. 1999b).

Although this argument has been framed in terms of leks, such mate choice considerations may also play a part in colony formation. If females preferentially chose males with nesting territories close to high quality individuals, this would allow them a greater chance of obtaining extra-pair fertilizations from the high quality neighbour (see Wagner *et al.* 2000 for a discussion of this idea). Under this hypothesis, there is a clear trade-off for these poorer quality males: although physical proximity to a high quality male may increase their chance of obtaining a mate, this is done at the cost of suffering a reduction in their probability of paternity.

A great number of theoretical models suggest that mate choice copying (where females are more likely to mate with a male that other females have previously selected) can be an effective strategy when sampling potential mates is costly or constrained, or if some females are better judges of male quality than others (e.g. Kirkpatrick and Dugatkin 1994; Dugatkin and Höglund 1995; Laland 1994a,b).

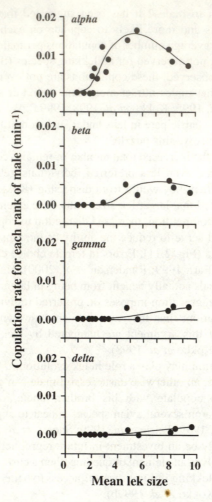

Fig. 2.11 Copulation rates for male ruffs of different ranks as a function of lek size. (Redrawn from Widemo and Owens 1995.)

There is also increasingly clear empirical evidence for non-independent choice of mates by females (Dugatkin 1992; Dugatkin and Godin 1992; Grant and Green 1995; Briggs *et al.* 1996). That said, not all studies that have looked for mate copying have found it (e.g. Lafleur *et al.* 1997), and some have reported that mate copying does not always override other preferences (e.g. Witte and Ryan 1998). Westneat *et al.* (2000) emphasized that non-independent mate choice can come about through a great many different mechanisms, several of which do not require elements of learning or cognition. There is a need for further empirical work to identify the mechanism that leads to non-independent mate choice by females, and a need for theory to address the evolution and fitness benefits of these different mechanisms.

2.7 Conserving heat and water

Animals can conserve heat by huddling together, because this will reduce the fraction of their surface area that is exposed to the colder surroundings. Assuming that the other individuals in the huddle are at the same temperature, no heat will be lost through the surfaces in contact between them, and this benefit should increase with group size (see Box 2.2). Substantial energetic savings through grouping have been demonstrated in several avian (Boix-Hinzen and Lovegrove 1998; Putaala *et al.* 1995; Ancel *et al.* 1997) and mammalian (Andrews and Belknap 1986; Bazin and MacArthur 1992) species, although counter-examples also exist (Berteaux *et al.* 1996). Beauchamp (1999), however, challenged the suggestion that thermoregulatory factors drove the evolution of communal roosting in birds. His argument was based on a comparative study, in which he found no evidence that communal roosting was associated with either high latitude or small body size. It may be that for many species, the thermal benefits of joining a roost are outweighed by the extra travel costs associated with flying to a roost site.

Another advantage to huddling in a confined space (like a burrow) is that heat lost from the bodies can significantly increase the heat of the surrounding air, thus reducing heat loss (Fig. 2.12; Box 2.2). The same will be true of water loss raising the

---

Box 2.2 Theoretical predictions of the energetic benefits of huddling

Theoretically, the extra gain that each individual receives from another joining the huddle decreases with increasing group size (Vickery and Millar 1984; Canals *et al.* 1989). Canals *et al.* (1997) suggested that huddling in a group of size *N* reduces the exposed surface area by a factor

$$\left(\frac{\phi}{N} + (1 - \phi)\right)^{0.735},$$

where the parameter ϕ is twice the surface area of animals in contact with other animals divided by the total surface area of all the animals. This theory was developed from the geometry of packing deformable spheres, with ϕ being a measure of deformability. This theory suggests that the maximum fraction of energy that can be saved by huddling is

$$1 - (1 - \phi)^{0.735}.$$

A follow-up paper (Canals *et al.* 1998) suggested that these predictions compared well with the limited amount of available data on several small mammals (Fig. 2.12), but further testing would be fruitful.

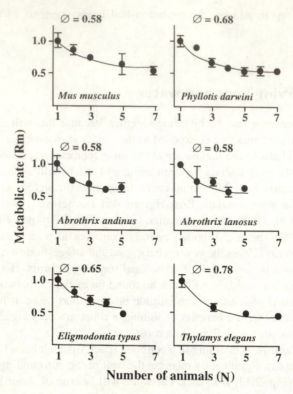

Fig. 2.12 The relative metabolic rates of different species of small mammals as a function of group size. The fitted line is the prediction of equation

$$\left(\frac{\phi}{N} + (1 - \phi)\right)^{0.735},$$

from Canals *et al.* (1997), with ϕ as a free parameter. (Redrawn from Canals *et al.* 1997.)

humidity. Hayes *et al.* (1992) suggested, on the basis of experiments, that this effect may be more important to small mammals than the effect of surface area reduction. Another important aspect to this change in the microclimate is that the animals that benefit most are the exposed ones on the periphery of the huddle, hence, in combination with surface area reduction, it should act to equalize the benefits of huddling between central and peripheral group members.

Huddling not only benefits endotherms: some slugs pack closely together with large areas of their flanks in contact, which has been shown to lead to reduced water loss through evaporation (Cook 1981). Clark and Faeth (1998) demonstrated experimentally that butterfly eggs in larger clusters were more protected from desiccation. The shape of the cluster was also shown to be important. They suggested that such water loss considerations may explain why egg clustering is more common in arid areas than in moist ones. Further evidence for this use of clustering to maintain water balance or body temperature in insect eggs or larvae can be found in Klok and Chown (1999) and references therein.

2.8 Reducing the energetic costs of movement

2.8.1 Introduction

The idea that an individual can reduce the amount of energy required to move at a given speed by placing themselves behind others, is familiar to human cyclists (Kyle 1979) and track athletes (Pugh 1971). That animals can obtain similar energetic savings has also been suggested by many theoretical studies and has received qualified empirical support.

2.8.2 Movement in water

Fish (1991) trained ducklings to swim behind a model of an adult inside a metabolic chamber, through which water flowed at a fixed rate. Chicks in a group of four, swimming behind a model adult suspended just above the water, reduced metabolic effort by 63% compared with a lone chick swimming behind the same raised adult. Although this reduction may be partially due to reduced stress levels of ducklings in groups, this effect alone is unlikely to explain such a dramatic change in energetic costs.

The spiny lobster (*Panulirus argus*) forms single file queues of over fifty individuals as they walk across the ocean floor on migrations. Bill and Herrnkind (1976) measured the force required to pull preserved specimens on a wire, and found that the force required to pull a line of individuals was considerably less that the sum of the forces required to the pull an equivalent number of single individuals at the same speed (Fig. 2.13).

Many fish swim as a group showing synchronized and polarized behaviour, which is generally termed schooling (Pitcher 1983). There now seems little doubt that when fish in a school are arranged in a certain spatial pattern, then individuals away from the front can gain considerable energetic savings, compared with swimming alone (Weihs 1973, 1975). Several authors have attempted to demonstrate that fish in groups use less oxygen (e.g. Abrahams and Colgan 1985, and references therein). However, Herskin and Steffenson (1998) argued that direct respirometry measurements of groups of fish are very difficult and the results of such studies hard to interpret. Further, as Pitcher and Parrish (1993) suggested, it is difficult to control for the possibility that fish in different group sizes may be engaged in different activities or may be in a different physiological state, although steps can be taken to minimize and quantify this problem (Abrahams and Colgan 1985).

Another line of evidence of energetic benefits to schooling would be to test whether spacing in natural shoals was consistent with the arrangement predicted by

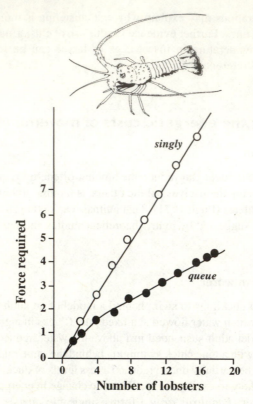

Fig. 2.13 Comparison between the total drag of a number of preserved lobsters towed singly or in a queue. (Adapted from Bill and Herrnkind 1976.)

the theory of Weihs (1975) to give maximum hydrodynamic benefit. Pitcher and Parrish (1993) reviewed the available evidence and found very poor support for the theoretical predictions. Abrahams and Colgan (1985, 1987) argued that the optimal arrangement from a hydrodynamic viewpoint is a poor one for anti-predator behaviour, because it would not allow fish to maximize the likelihood of predator detection. They further claimed that increasing the predation risk caused groups of fish to reorganize themselves in a way that made the group less hydrodynamically efficient, but which was likely to reduce predation risk. A functional interpretation of two- and especially three-dimensional group structures will need careful consideration of the assumptions underlying Weihs' theory and remains a challenge for future studies.

There is good evidence that a fish's tail beat frequency is related to its energetic expenditure during travel (Herskin and Steffensen 1998), and several studies report that fish at the back of a school beat their tails at a lower rate than those at the front (Herskin and Steffenson 1998; Fields 1990; Zuyev and Belyayev 1970). Thus, in conclusion, it seems likely that fish are able, at least under some circumstances, to make hydrodynamic savings through schooling behaviour. However, our understanding of this is far from complete. Further theoretical work, or perhaps physical models in the spirit of Bill and Herrnkind's (1976) experiment are urgently required. An interesting challenge for the theoretician would be to compare fish with squid, which form schools (Mather and O'Dor 1984), or salps (free-swimming tunicates), which form chains (Bone and Trueman 1983), but use completely different locomotive mechanisms to fish.

2.8.3 Movement in air

There is a well established theory that aircraft can save energy by flying in formation (Hummel 1995). This theory has been supported by experiments with fixed-wing aircraft (Hummel 1995). According to Lissaman and Shollenberger (1970), this theory can be transferred to birds using flapping flight provided the ratio of wing-tip speed to forward flight speed is low, a condition met by the slow flapping flight of large birds. Hence, energy saving has been suggested as an explanation for the formation flight of some large birds species, particularly on long distance migrations. It is thought that the wake patterns produced by smaller birds are too complex and variable for another bird to be able to fly in a position that would allow energetic savings. Hence it is unlikely that this energy saving mechanism applies to small passerines (Hummel 1995). This view is strengthened by the argument of Higdon and Corrsin (1978), on theoretical grounds, that improved flight efficiency is not an important factor in three-dimensional flocks, such as displayed by roosting starlings.

The theory for larger birds has been well developed (Higdon and Corrsin 1978; Hummel 1983; Hainsworth 1987, 1988) and the maximum saving is achieved when birds are in a staggered formation with their wing-tips overlapping in the direction of flight, as would be achieved by a V-shaped formation. However, observation of naturally occurring formations found that overlap was less than theory predicted (Hainsworth 1987; Cutts and Speakman 1994; Speakman and Banks 1998). This can be explained on the grounds that there are severe energetic costs to overlaps greater than the optimal, so birds err on the side of caution by adopting a positioning slightly less than optimal, to avoid wind (or other factors) moving them into the high-expense zone. Badgerow and Hainsworth (1981) reported spacings close to the optimum in Canada geese, although tending towards excessive overlap.

Much of the theory has concentrated on symmetric V- or asymmetric J-shaped formations. However, these make up the minority of observed shapes, with simple staggered lines with no apex being most common (Gould and Heppner 1974). We do not know of any mechanism to explain why staggered line formations are so common.

An alternative or complementary theory explaining group formation among migrating birds is communication of navigational information. This theory predicts that wing-tip spacing and depth (perpendicular to the direction of travel) should be correlated within a species, as individuals attempt to keep the bird in front in the centre of their visual field (Gould and Heppner 1974; Badgerow 1988). Such a relationship, albeit a very weak one, was found for pink-footed geese by Cutts and Speakman (1994), although not in Greylag geese using the same methods (Speakman and Banks 1998). However, Heppner *et al.* (1985) argued that Canada geese typically do not fly at the angle that their visual physiology would predict. Effective testing of the relative importance of navigation and energetic savings to migrating birds remains to be done. Another underexplored consideration is the role of flight speed for grouping: Kshatriya and Blake (1992) argued that optimum flight speed decreases with group size. This awaits both empirical testing and further theoretical consideration as to whether this sets an optimal size for flocks.

Testing of the theory that birds actually experience an energetic benefit from formation flight has only recently been obtained by Weimerskirch *et al.* (2001). These authors found that pelicans' heart rate was 14% lower when flying in a V-shaped group than when flying alone.

2.9 Summary and conclusions

Our current understanding of the mechanisms that can lead to benefits from grouping seems quite well advanced. This stands in contrast to several of the aspects of grouping that we will discuss in coming chapters. That said, there are still some obvious gaps in our understanding that need to be filled.

It seems clear that there can be real anti-predatory gains to grouping. However, empirical work on this has been heavily biased towards birds. An exploration of how our understanding from avian systems can be extrapolated taxonomically would be very useful. Even for avian systems, some very basic assumptions of the theory remain untested. For example, there seems to be little evidence that reduced individual vigilance allows an individual to increase time spent feeding or engaged in some other advantageous activity.

Similarly it now seems very clear that foragers can gain information from the foraging activities of others, and that this information can be used to improve foraging efficiency. Here too there is still some room for useful further development. Current theory makes assumptions about the type of information that can be collected, its reliability and accuracy, and about the costs of acquiring the information. However, these assumptions have generally not been exposed to careful empirical testing. Consideration of limitations to the quality and quantity of information that can be gained from others, and how this can be integrated with information from the focal individual's own experience, should considerably increase our understanding of this mechanism.

Lastly, it seems that the savings in transport costs made possible by co-ordinated group movement have been much more extensively studied in air than in water. This bias almost certainly derives from challenges in data collection rather than being a reflection of the relative importance of the mechanism in the two media. The recent demonstration that group formation can reduce the oxygen consumption of krill by a factor of seven (Ritz 2000) shows the very considerable savings that can be made. We hope that such spectacular results will drive both empiricists and theoreticians to take a fresh look at the energetic consequences of fish schooling.

3

Some costs to grouping

3.1 Introduction

Here, we consider the down-side of grouping, mechanisms that can lead to decreased fitness with increasing group size (see Box 3.1 for a summary). It is important to realize that the main fitness consequence of a potential cost, may not be that of paying the cost itself, but of the effort or constraint involved in avoiding this cost. For example, in gull colonies, egg and chick predation by other colony members can be an ever-present threat. Some parents will experience this cost directly, but some will not. However, all parents (whether their eggs suffer predation or not) will pay the costs of trying to minimize the risk of such cannibalism. This may be seen as a constraint on foraging, as at least one parent must remain on the nest at all times. Alternatively or additionally, there may be an energetic cost due to aggressive

---

Box 3.1   A summary of the potential disadvantages of grouping

(a) Simple competition, the more individuals that share a finite resource, the less each will get.
(b) Food discovered by one animal but then stolen by another (kleptoparasitism) should generally lead to a cost being experienced by at least some group members.
(c) Aggression between group members for reasons unrelated to foraging.
(d) Pseudo-interference. This term is commonly used for the reduction in foraging efficiency suffered by individuals that are forced by competition to use intrinsically poorer feeding sites.
(e) For sessile foragers, the path that mobile food takes to them can be blocked by the presence of another forager.
(f) Individuals can simply get in each other's way.
(g) Prey may more easily detect an approaching group than an individual and take more effective evasive action.
(h) Predators may preferentially target larger groups of prey.
(i) Confusion or interference between fleeing individuals may increase the success rate of predators with increasing prey group size.
(j) Area restricted search tactics by predators are particularly effective against grouped motionless prey.
(k) Greater proximity between breeders may increase the risk of misdirected parental care.
(l) Increased proximity of individuals is likely to increase the likelihood of transfer of pathogens.

interactions required to defend the nest site from conspecifics. More subtly, temporal synchrony of breeding may serve to reduce the opportunity for cannibalism, but it may also increase competition for food.

As in the last chapter, we begin by considering the consequences of grouping on risk of predation. It may be that groups of prey are more easily detected by predators than single individuals and so suffer more attacks. This will certainly act to diminish the anti-predatory benefits of grouping discussed in the last chapter. An extreme case may be the cost to shoaling in oceanic fish; because the sonar used by predatory cetaceans (and human fishing boats) cannot detect single fish but readily identifies aggregations. Isolated individuals should be virtually invulnerable to these predatory threats, but grouping (selected for other reasons) may bring with it a cost in increased predation. Section 3.2 looks at the empirical evidence for larger groups suffering higher rates of predatory attack than small ones. Surprisingly, for such a conceptually simple but fundamental idea, our understanding is still relatively poor. Many empirical studies fail to control for confounding variables, and few, if any, are able to demonstrate that predators have a preference for larger groups, and to quantify how this preference varies with group size. There are fragments of theory for how detectability should change with group size, but these ideas have remained largely untested.

Section 3.3 concentrates on cataloguing the many different mechanisms that can lead to a decrease in foraging efficiency with increasing group size (see Box 3.1). These mechanisms are not exclusive, and we would expect that most foragers suffer from at least one. There can be little doubt that large groups foster disease transmission and Section 3.4 documents the evidence for this. Similarly, the likelihood of ending up investing in another individual's offspring is higher for group-living individuals which is discussed in Section 3.5.

3.2 Increased attack rate on larger groups

It seems intuitive that larger groups will be easier to detect; this intuition is supported by a limited amount of theory (Box 3.2). The relationship between predation rate and group size has been investigated in several field studies involving raptors attacking flocks of avian prey. Cresswell (1994) studied naturally occurring attacks by sparrowhawks on redshank and concluded that larger flocks were preferentially attacked. This occurred despite attacks being more likely to succeed against smaller flocks (Fig. 3.1). Whether this preference occurred because group size was non-randomly associated with time of day, time of year, or location is not discussed. A similar caveat applies to the study of Szép and Barta (1992), who observed a positive correlation between the number of swallows flying around a colony and the rate of attack by hobbies. Lindström (1989) found that both the rate and the probability of success of sparrowhawk attacks on finch flocks increased with flock size. He observed that 'the largest flocks almost always foraged under exposed conditions', which suggests the group size may well have been non-randomly associated with some other relevant factor, and that sparrowhawks need not be targeting larger

Box 3.2 Theoretical considerations of how detectability might change with group size

Vine (1973), using statistical fitting to data from humans, argued that the detection range of a line of n similar targets viewed from the side increases as n to the power B where $0.4 < B < 0.45$. For targets tightly bunched into a circle, he suggested $0.2 < B < 0.23$. A useful extension to this theory would be to consider detection from angles other than that obtained by viewing a planar shape from the same level. For example, groups viewed from above by flying animals have not been considered. Whilst further empirical testing of these relationships would be interesting, they may be limited in their usefulness because they are not based on any underlying theory that can shed light on the mechanisms of detection.

Kunin (1999) suggested that detection range by olfaction should increase as n to the power 1, based on the physics of diffusion. Whilst this may be true of olfactory detection, similar rules are unlikely to apply to different sensory systems. A key question that must be addressed in the further development of this theory is empirical verification of the assumed mode of information transfer. For example, diffusion may be the relevant transport process for some olfactory detection, but bulk fluid motion is likely to dominate in most systems.

It would seem that the most effective way forward after identification of the key physical processes would be the development of a mechanistic model based on our understanding of these processes and of the physiology of the sensory organs involved. A good example of this is the work of Asknes and Utne (1997), which is based on a complex model of the optical properties of fish eyes, which suggested that the visual range r through water at which a target of area A can be detected is given by

$$r^2\exp(-cr) = kA,$$

where c and k are constants. Given that the visual area of a group will generally change in a non-linear way with group size, the effect of group size on detectability may be quite complicated. We clearly need an expanded body of theory that can be exposed to rigorous testing.

groups *per se*. He also suggested that the higher success rate may be because larger groups are more likely to contain an injured or poorly conditioned bird. Page and Whitacre (1975) reported cold weather as a confounding factor, leading both to an increase in attacks by Marsh hawks and kestrels, and to a change in flocking behaviour in their dunlin prey. In contrast, Trail (1987) reported that 'raptor attack rate was inversely correlated with group size', in lekking Cock-of-the-Rock. That is, the larger a group, the less frequently it was attacked. However, the most likely explanation for this is that times of low raptor attack rates (for whatever reason) allowed large leks to build up.

Treherne and Foster (1982) observed natural predation by fish on flotillas of marine insects and found that the rate of attacks on a flotilla was independent of its size, despite attacks being more successful on smaller groups. FitzGibbon (1990) observed that cheetah preferentially attacked smaller gazelle herds, perhaps reflecting their great success rate in attacks on smaller herds. In conclusion, while there is a great deal of evidence for a change in attack rate with group size, the reasons for this are unclear, and there is certainly not strong evidence that predators

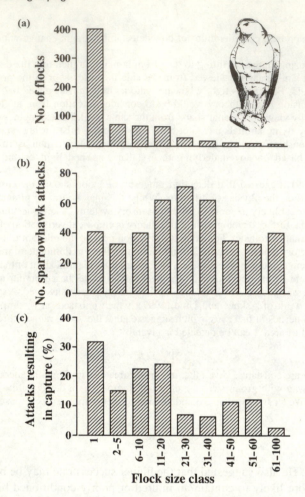

Fig. 3.1 Histograms showing (a) the distribution of redshank flock sizes, and (b) how attacks by sparrowhawks were spread between redshank flocks of different sizes, in the study of Cresswell (1994). (c) Probability of an attack being successful changed with flock size.

preferentially target larger groups when potential confounding factors are eliminated or accounted for. In general, the studies above were not designed to focus on this question, and further work is required.

There is evidence from laboratory experiments that fish predators showed a preference for larger groups when offered a choice between two groups, differing only in size, arranged side by side in an aquarium (Krause and Godin 1995) (Fig. 3.2). Two interesting additions to the latter observation are that predatory attempts were less successful against larger groups, and that predators could be made to preferentially attack the smaller group if fish in that group were more active than the other group (Fig. 3.2). The authors suggested that this may be because the more active group was more visually conspicuous to the predator. It may be that the predator

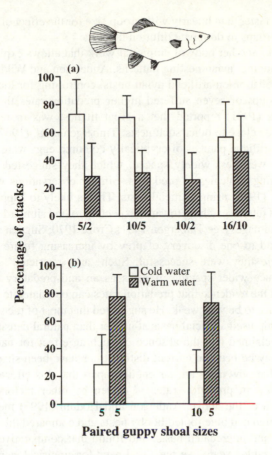

Fig. 3.2 In choice trials performed by Krause and Godin (1995), the percentage of attacks directed at guppy shoals of different sizes when a predatory cichlid fish was offered a choice between two shoals (a) differing only in size, or (b) differing in the temperature of water that they were held in or in both size and temperature. Data are presented as means ± standard deviations.

preference for more active groups is adaptive, if activity tends to be related to foraging behaviour that results in reduced anti-predatory vigilance, and hence increased vulnerability to predators. Riipi *et al.* (2001) using great tits trained to search for novel objects in an aviary against a similar background found prey groups of four and eight prey to have a higher detectability risk than solitary individuals, although the increase was not proportional to group size.

As discussed in Chapter 2, such a preference for group size must be interpreted within the general framework of attack abatement (see Section 2.3.1). Thus, we need to understand not just whether predators have a preference for larger group sizes, but how this preference varies with group size. For a predator that can catch at most one individual in a given attack, then (all other things being equal) the rate of attacks will have to increase faster than linearly with group size for the effect of the higher attack rates on larger groups to dominate dilution benefits.

We now turn to another body of empirical work that allows exploration of predation costs to grouping: avian nesting patterns. Andersson and Wiklund (1978) compared natural predation on artificial avian nests, controlling for location, and found that nests in groups of seven suffered higher predation rates than solitary nests. Similarly, Krebs (1971) reported that a great tit nest was more likely to suffer predation if it was close to other such nests. Tinbergen *et al.* (1967) found that predation rates on artificial patches of cryptically coloured eggs were higher when the eggs in a patch were less widely spaced, which they suggested was due to area restricted searching strategies of predators. Similar observations were obtained by Göransson *et al.* (1975) using artificial nests. This is likely to happen because predators tend to bias future searching towards expecting to find similar items (i.e. forming a specific search image) (see Tinbergen 1951; Croze 1970; Sugden and Betersbergen 1986), or respond to one discovery of prey by increasing future searching in the vicinity of where they were successful. Such 'area restricted' search would be expected to induce wider spacing of nests as an anti-predatory defence. Andren (1991) reviewed the evidence that predation rates can be related to nest spacing and found the evidence to be very weak. He suggested that most of the studies that found such a relationship used artificial nests at higher than natural densities. Schieck and Hannon (1993) claimed that the absence of such an effect for naturally occurring nest densities may be because current distributions have been shaped by predation pressure, such that now nests are far enough apart that this effect is not seen, and therefore variation in predation rates is driven by other factors. As a counter-example to much of the above, Anderson and Hodum (1993) measured the local density around predated blue footed booby nests and compared this to the local density of nests around unpredated nests, and found that comparatively isolated nests were more vulnerable. Again, we see a real need for empirical work into the search strategies adopted by predators whose prey tends to be aggregated but cryptic, such that the whole group is not discovered at once. At present, there is no strong evidence of a general increase in predation risk with increased grouping of nests.

As a last thought on predation costs to grouping, it is worth remembering that predation can also be intra-specific. Moller (1987) reported that larger swallow colonies suffered more heavily from infanticide by unmated males, in an attempt to induce re-mating, than smaller colonies. This may be because such individuals are more tolerated in larger colonies because the large numbers of nesting individuals makes identification of non-nesting ones more difficult. Similarly, colonial sea-birds are particularly prone to suffer from egg and chick cannibalism (Wittenberger and Hunt 1985).

3.3 Foraging in a group

It is clear that when a food patch is found, or a prey item captured, the more individuals that share this resource the smaller the per capita share. This cost to group foraging will almost always apply. One of the few exceptions to this may be for situations, such as sea-birds feeding on a fish shoal near the surface, where the resource is so bountiful that depletion by others does not measurably decrease the food available to any one individual. However, there are many other mechanisms by which conspecifics can adversely affect the feeding of others. Such depression of individual instantaneous feeding rates as a result of the proximity of others is generally termed interference.

3.3.1 Kleptoparasitism

Group foraging exposes an individual to the possibility of food items being stolen by another group member, before that individual has had a chance to process the item. This stealing of resources is often termed kleptoparasitism. Some individuals can also gain by this, if they are able to steal more items than are stolen from them, conversely this means that some individuals lose out. Further, attempting kleptoparasitism will have some cost to both parties. This may be measured in terms of time or energy that must be devoted to an aggressive interaction, the risk of injury or worse, or because vigilance for kleptoparasitic opportunities can only come at the expense of reduced ability to find food by conventional means. Kleptoparasitic attempts may also result in the escape, loss, damage, or destruction of the prey item. These costs and benefits mean that we expect kleptoparasitism only to be used by individuals when it is advantageous to them. The cost/benefit trade-off will depend on the characteristics of the prey item in question and the two potential combatants for it, as well as the availability of other prey in the environment. The idea that individuals should show behavioural plasticity in kleptoparasitism, and that the behaviour should only be undertaken by individuals that would benefit from it, has only recently been considered by theoreticians (Broom and Ruxton 1997; Stillman *et al.* 1997; Sirot 2000). This body of theory makes very clear predictions about the circumstances under which kleptoparasitism should be seen and how this behaviour leads to an interference effect of reduced intake rate as group size increases. These results are very much in need of empirical testing. Dolman's (1995) study of small passerine birds and the extensive work that has been carried out on shore-birds—particularly the oystercatcher (e.g. Stillman *et al.* 1996)—provide good templates for such work.

3.3.2 Aggression more generally

There may be circumstances when aggression occurs between group members for reasons other than foraging. This might occur, for example, when the purpose of aggression is related to the maintenance of dominance hierarchies, or some such mechanism not related directly to foraging. Under such circumstances, interference will still occur, but this can often be addressed by simpler models than those considered in the last section, since we need not consider the short-term net-value of aggressive behaviours, in order to evaluate its effect on foraging. Rather, we can safely consider aggressive behaviour to be inflexible, with an individual either always or never being aggressive regardless of circumstances. The effect of such behaviours on the interference experienced by an individual has been addressed by theoretical models for some time, the most relevant works being Ruxton *et al.* (1992) and Holmgren (1995).

3.3.3 Pseudo-interference

The size of a group of foragers can have an adverse effect on individual foraging even when no aggressive interactions are apparent. We have already discussed the idea that larger numbers of individuals will generally deplete a resource more quickly, with each individual getting a smaller share. This depletion can cause the individuals to exploit less profitable feeding opportunities, that would not be exploited if competition were less intense. This use of poor quality feeding options in turn leads to a reduction in feeding rate with increased group size, termed pseudo-interference by Free *et al.* (1977). Notice that the above mechanism does not require animals to make decisions based explicitly on group size. However, it may be that individuals do sometimes actively make foraging decisions based on a consideration of competitor numbers *per se*. For example, individuals might be slower to settle in an area that already contains competitors. This can lead to a strengthening of the pseudo-interference effect, termed mutual indirect interference by Driessen and Visser (1997). Although there is some controversy over the exact definition of this term, and its relation to pseudo-interference (see Weisser *et al.* 1997; Lynch 1998), there is certainly a need to evaluate the importance of these more subtle forms of interference.

In a simple system of birds feeding on the ground, Cresswell (1998) observed that the feeding rates of birds were depressed when foraging in proximity to others, even when no aggressive interactions occurred and prey depletion was unimportant. He discussed two (potentially complementary) mechanisms where the threat of potential aggression produces this effect. First, investment in watching a potential aggressor perhaps can only be achieved at the price of reduced investment in foraging. Secondly, birds may modify their movement across the ground in order to avoid a potential aggressor, such that the probability of covering previously searched (and harvested) ground is increased.

3.3.4 Shadow interference of sit and wait predators

For sit and wait predators, including filter feeders, individuals that place themselves in close proximity to another interfere with this individual's foraging by intercepting food that otherwise would have reached the focal individual. This has been termed 'shadowing', since one individual suffers from being in the 'shadow' of another (Wilson 1974). Minimization of this type of interference has been suggested to lead to the doughnut-shaped aggregations often observed in the pits that antlion larvae use to trap crawling insects (Linton *et al.* 1991). Shadow interference has also been suggested to cause aggression to be biased towards upstream competitors in stream living filter feeding insects (Hildrew and Townsend 1980). As well as reduction in

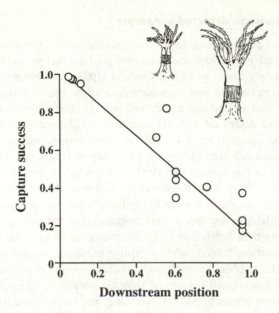

Fig. 3.3 The density of particles available to filter feeding polyps in a colony of boreal octocoral (normalized to the density at the upstream extreme of the group of polyps), plotted against the position of a polyp relative to the upstream edge of the group. (Adapted from Patterson 1984.)

prey availability through direct capture, such 'upstream' competitors may cause a further interference effect through alerting prey to potential danger (see Section 3.3.6) or through modification of the local flow conditions in freshwater systems. Patterson (1984) demonstrated a contrasting situation for passive suspension feeders in water. He found that large aggregations affected water flow patterns in such a way that greater mixing could occur, increasing the amount of food available to group members (Fig. 3.3).

3.3.5 Just getting in each other's way

Imagine a group of N foragers that exhaust a food patch and then leave, searching independently for another food patch. Their searching ability is likely to be better than that of a single individual, but by a factor of less than N. The reason for this is that the areas they search are likely to overlap, since they start from the same point in space. The effect of this overlap has been investigated theoretically by Ruxton (1993), but much remains to be done to explore the importance of this factor. Such overlap will be affected by the search patterns of individuals, and particularly by how these search patterns are influenced by the relative positions of other group members. Further, a trade-off exists between being as far from other group members as possible to minimize the risk of re-searching ground already searched by others and being as close as possible in order to take maximum advantage from sharing the finds of others.

3.3.6 *Prey response to detected predators*

Another neglected aspect of interference is the reduction in the prey available to an individual due to the prey's response to another predator that passed by recently. This was investigated by Selman and Goss-Custard (1988) who observed wading birds (redshanks) feeding on mud flats. A square area of the flats was delineated and the time taken by a bird to find three prey individuals after entering this square was recorded. This time decreased with the time that had elapsed since the last bird had passed across the square (Fig. 3.4). That is, the more recently a conspecific had passed through the study area, the poorer the foraging conditions were. Prey is sufficiently plentiful in this system that depletion can be dismissed as an alternative explanation for this effect. Also, there was no indication that the birds' foraging rate was affected by the distance between individuals. Thus, the authors concluded that reduced prey availability was due to prey responding to the passage of the previous bird by modifying their behaviour for a short time so as to make themselves less vulnerable to predation. Indeed, when a captive redshank was made to walk across a patch of mud, the numbers of amphipod crustacea coming to the surface decreased sharply and stayed depressed for several minutes afterward. Again, this mechanism requires much more research, and will certainly not be confined to redshanks and

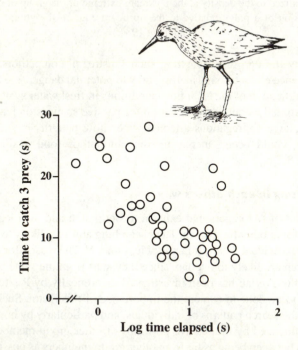

Fig. 3.4 The time taken by a redshank to catch three prey items as a function of the time elapsed since another bird fed in the same place. (Adapted from Selman and Goss-Custard 1988.)

their prey. A theoretical framework for studying this phenomenon has recently been introduced by Stillman *et al.* (2000). This model demonstrates that foragers can greatly reduce the impact of prey responses by modifying their search paths, using information on the foraging behaviour of others. The authors argue that just such a behavioural modification appears to be performed by oystercatchers.

3.3.7 A note on generality

Although all the forgoing has discussed the pursuit of food, many of these ideas will translate directly to animals in groups pursuing other resources. For example, as we discussed in Section 2.6, one disadvantage to communal displays by males seeking females is that copulation attempts are often interrupted by other males seeking to maximize female availability to themselves.

3.4 Increased parasite burdens

Higher levels of contact-transmitted parasites have been found to be associated with larger groups in mammals (Hoogland 1979; Van Vuren 1996) and birds (Brown and Brown 1986; Hoogland and Sherman 1976), but not for lizards (Sorci *et al.* 1997). Hoi *et al.* (1998) found that infestation with parasites in one species of colonial nesting bird (the bee-eater) decreased with increasing inter-nest distance. Comparison between two closely related bird species that differ in their social structure (territorial hooded crows and colonial rooks) showed that they share many parasites, but that infestation was greater in the rooks (Rozsa *et al.* 1996). One comparative study found that group-living passerine birds had a greater infestation of feather mites than sympatric solitary species (Poulin 1991), but another found no effect (Poiani 1992). It seems likely that differences in ecology between species make interspecific comparisons difficult, however it may now be fruitful to revisit this question after a further decade of data gathering and methodological advance.

Moore *et al.* (1988) looked at parasite loads in another bird, the bobwhite quail. They found a positive relation between load and group size only for monoxenous (single host species) parasites with a short life cycle. Brown and Brown (1986) demonstrated a very strong cost of parasitism in cliff swallows (see Fig. 3.5), so there can be no doubt that parasitism can exert an important selection pressure. In fish, Poulin (1999) found that shoals of sticklebacks inhabiting tidal pools with no water current had a positive relationship between shoal size and infection levels by a copepod, which is transmitted among fish by short-lived planktonic larvae. Côté and Poulin (1995) reported that a meta-analysis of empirical results shows a consistent positive correlation between group size and both the prevalence and intensity of contagious parasites. Taken together, these studies point to a general trend: group-living produces an increased risk of infection from diseases and parasites that

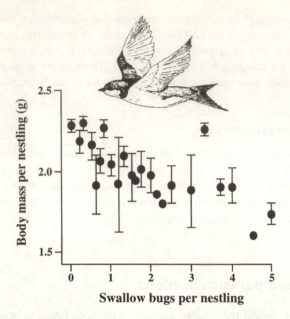

Fig. 3.5 The relationship between degree of parasitism (mean ± S.E.) and nestling body mass in the study of Brown and Brown (1986).

rely on simple short-range (often undirected) vectors, as opposed to the more behaviourally flexible parasites described in Section 2.3.4.

However, it is possible that group formation could act to reduce parasite loads in some situations, by allowing allogrooming (Loehle 1995). Alternatively, a territorial social system could lead to increased aggressive interactions where bite wounds make transmission of some diseases more likely (Loehle 1995).

3.5 Misdirected parental care

3.5.1 Cuckoldry

This cost is only of concern to males that provide parental care, and indeed may hold fitness advantages for females, as detailed by Petrie and Kempenaers (1998). Moller and Birkhead's (1993) comparative study concluded that colonially breeding birds experienced higher rates of extra-pair copulations. However Westneat and Sherman (1997), in another interspecific comparison, did not find that extra-pair fertilizations increased with density (see Fig. 3.6). These authors discuss potential biases in the measurement of copulations and confounding effects inherent in interspecific studies, and report that 8 out of 11 studies using intraspecific comparisons found that

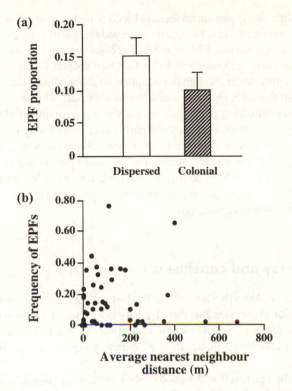

Fig. 3.6 (a) The average (± standard deviation) proportions of chicks sired by extra-pair fertilizations in both dispersed and colonial breeders (from a review by Westneat and Sherman 1997). (b) The frequency of extra-pair fertilizations as a function of the characteristic distance between nests (from the same study).

extra-pair fertilizations increased with breeding density. It seems likely that physical proximity of protagonists will have a positive effect on the ease with which extra-pair copulations and fertilizations can be achieved. However, a general understanding of the relationship between colony size and/or density and risk has yet to be achieved.

3.5.2 Brood parasitism and adoption

In Section 2.5, we discussed experimental evidence that colonial breeding allows collective defence against brood parasites and predators. However, there may be associated costs to coloniality, as this may allow a brood parasite to simultaneously observe a large number of nests, either to evaluate the quality of the parents or to time parasitic acts to parental non-attendance. These factors may be particularly important for intraspecific brood parasitism, where parasitic individuals often also raise a brood themselves. It appears that a disproportionate number of species that show intra-specific brood parasitism seem to breed at high density, however this has not been rigorously explored. Yom-Tov (2001) surveyed reports of avian intraspecific brood parasitism and suggested that 57% of the 232 species reported to show this behaviour were colonial breeders, whereas only 13% of all bird species breed colonially.

Another cost may occur through the adoption of avian chicks that hatched in their parental nest but then subsequently moved to another nest. This may happen by accident, particularly following a disturbance, but there is growing evidence that chicks that are being poorly provisioned by their parents are more likely to attempt to move to another nest (see Davies 2000 for an overview). Since the world outside a nest is generally dangerous for chicks, such adoption attempts would only be possible where breeding occurs at high density. Compared, for example, to the huge interest that is devoted to alloparental care by 'helpers' at the nest (see Section 2.1), adoption has been notably understudied.

3.6 Summary and conclusions

This chapter is considerably shorter than the last one, which explored the benefits of grouping. It is our impression that this disparity is a result of a bias in scientific activity. The lack of study is reflected in the large proportion of sections in this chapter where we have concluded that current evidence is equivocal and further research is needed.

In our view the least well understood cost of grouping is that attack rates by predators may increase with group size. Since many of the benefits of grouping relate to mechanisms for reducing predation risk, this deficiency in our knowledge is a serious one. Natural predation events are challenging to study, being generally hard to predict in space and time. Thus, ease of data collection has probably been one of the main driving forces behind a research tradition that has focussed more on the prey than the predator. There is a great imbalance between our considerable understanding of vigilance patterns of prey species compared to a very scanty consideration of the factors that govern target choice by predators. Methodological challenges notwithstanding, we must learn more about prey choice by predators. This should involve laboratory experiments as well as fieldwork. For example, the theory of how detectability changes with group size (discussed in Box 3.2) should be very amenable to testing under controlled conditions.

For predators, and for searchers more generally, there has been considerable empirical and theoretical consideration of how search paths are influenced by prey discoveries, but very little on how these paths are influenced by other searchers in the vicinity. Given the diversity of interference mechanisms (as well as the potential benefits of local enhancement), that depend critically on the relative positions of individuals, there is a real need for our knowledge to be expanded in this direction.

In summary, we don't know nearly as much about the costs of grouping as the benefits, pick any topic you like, but go out and close this gap!

4

The size of a group

4.1 Introduction

4.1.1 Combining costs and benefits of grouping

The last two chapters discussed the mechanisms by which being in a group can affect an individual. Of course, these mechanisms are often not mutually exclusive, and being in a group will affect an individual in several different ways simultaneously. Further, the different costs or benefits that an individual derives from these different mechanisms will change differentially with changes in group size. Section 4.1 will outline the idea that this differential change in costs and benefits will often lead to the existence of a unique group size that allows individuals to maximize fitness, the 'optimal group size'. Deviations from this group size will be disadvantageous to group members.

However, we will then go on to argue in Section 4.2 that there is reason to expect that group sizes found in nature will generally be greater than the expected optimal. How much greater will depend on such things as the relatedness between individuals, the existence of dominance relationships, the ability of individuals to change between groups, whether or not groups can exclude individuals, and the degree to which individuals can co-ordinate movements. This section will be largely theoretical, but we will emphasize that recent developments of this theory make predictions that should be amenable to testing. One important consideration is that theoretical predictions depend strongly on whether the decision to expand a group is taken solely by the potential joiner or is controlled by those already in the group.

One would expect to observe active recruitment to groups that are of a size below the optimal one. However, evidence for active recruitment to foraging groups is scant and we outline how future studies can provide more powerful evidence. We further argue that there is a need for theory that considers ecologically more realistic scenarios giving individuals a choice between joining groups of different sizes, rather than a choice of joining a given group or foraging alone.

Empirical testing of optimal group size theory has been even more rare than that of active recruitment theory. This is surprising given the popularity of the paradigm. In Section 4.3, we outline the data that is available, and argue that the problem with previous studies is that by choosing to focus on large mammals, students of optimal group size theory have picked systems where data collection is very difficult and experimental manipulation near-impossible. Hence, we argue for a change of study system, suggesting social spiders as a promising test-bed. We finish this chapter by reviewing some of the available data on the distribution of group sizes in free-ranging species from different taxonomic groups, such as African buffalo and pelagic fish. To date, theory has focussed on producing models that can mimic these distributions. To our mind, insufficient attention has been given to understanding the mechanisms by which these distributions arise.

4.1.2 An illustrative example

Imagine a bird, searching for invertebrates on a lawn, which is joined by another bird. This might impact on the original bird through reducing its predation risk. There are now four eyes to scan for predators instead of two, and even if an attack is not detected in time, the probability that the original individual will be the focus of that attack may have dropped from 100% to 50%. There may, however, be a disadvantage to being joined, in terms of reduced feeding rate. Perhaps through prey response or through modification of search paths so as to avoid the other bird, there may be an interference effect of the new bird joining. However, in many circumstances, the anti-predatory advantage of another individual arriving will outweigh this disadvantage.

Imagine now that a third individual arrives, this individual too will bring an anti-predatory advantage. However the incremental benefit to our original focal individual is likely to be less than when the previous individual arrived. For example, the risk of being the focus of a randomly targeted attack falls from 50% to 33%, a 17% decrease, rather than the previously experienced 50% decrease. For every further new arrival, we would expect that the additional anti-predatory benefit to the focal individual becomes less and less (Fig. 4.1a). However, we would not generally expect the additional foraging disadvantage to decrease in the same way. Indeed, we might perhaps find that the interference effect of two other individuals is more than twice as great as that of one other individual. Thus, as more and more individuals are added, we find that the added net advantage to the focal individual becomes less and less. Indeed, after a certain group size has been attained, the disadvantages begin to outweigh the advantages, and further arrivals lead to a net disadvantage to the focal individual. In such circumstances, we predict that there will be an optimal group size where the net benefits to group members are maximized. Any deviation from this group size is disadvantageous (Fig. 4.1b).

4.1.3 The shape of the fitness function

The situation described above (with a single optimum group size) is likely to be common in nature, but need not always apply. In some circumstances, it may be that grouping is always disadvantageous and individuals maximize their fitness by being alone (such a fitness function is shown in Fig. 4.2a). However, we cannot think of circumstances where increasing the size of a group always brings net benefits, no

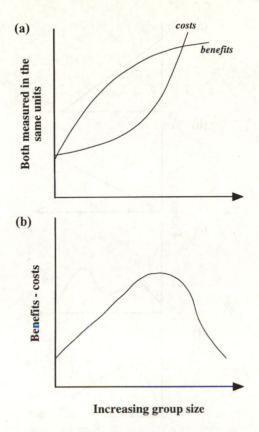

Increasing group size

Fig. 4.1 (a) The costs and benefits to an individual of being in groups of different sizes. As group size increases, both costs and benefits increase. However, the increase in benefits is a decelerating function (with each added individual having less effect than the last), whereas the increase in costs accelerates (with each added individual having more effect than the last). This means that the costs will eventually outweigh the benefits for large group sizes. (b) The difference between the costs and benefits as a function of group size. We see that this gives a characteristically *n*-shaped graph, with a single maximum at some intermediate group size (the so-called optimal group size).

matter how large the group becomes (Fig. 4.2b). Eventually, if the group becomes large enough, then the costs of grouping will dominate. It is, however, possible to construct circumstances where fitness functions have more complicated shapes, with more than one turning point (Fig. 4.2c). Although in most circumstances where grouping is advantageous at low group sizes, we would expect there to be a unique group size that maximizes individual fitness (Fig. 4.2d). The next section considers whether we should expect to see such optimal group sizes in nature.

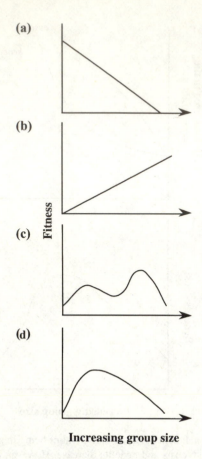

Fig. 4.2 Four alternatives for the relationship between individual fitness and group size. (a) Fitness always decreases with increasing group size. (b) Fitness always increases with increasing group size. (c) A fitness function with more than one optimal group size. (d) An *n*-shaped graph, like Fig. 4.1b, with a single maximum at a unique optimal group size.

4.2 Are optimal group sizes likely to be seen in nature?

4.2.1 An argument why groups should be larger than optimal

Sibly (1983) was the first to question whether groups of optimal size were stable. Consider a series of individuals arriving at a site singly. After the first individual, each has a choice between foraging on their own or joining a group. We assume that each individual's fitness changes with group size as described by Fig. 4.3. We can see that the second individual that arrives does better by joining the other individual and being in a group of size two, rather than being alone. The same argument holds for each individual that arrives until we reach the optimal group size of 20.

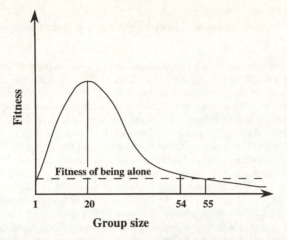

Fig. 4.3 A fitness function with the following properties. Fitness has a positive non-zero value for a lone individual (in a group size of one), it then increases with increasing group size, until it reaches a maximum at the optimal group size of 20. After that, it declines for further additions to the group. When the group size reaches 54, the fitness of individuals in the group is just slightly greater than that of a lone individual. However, if the group size reaches 55, then individual fitness is equal to that of a lone individual. For group sizes larger than 55, individual fitness continues to decline. Although the fitness function itself is not symmetrical, it is approximately so around the optimal group size. (Adapted from Sibly 1983.)

Now consider what happens when the 21st bird arrives. It has a choice between feeding in a flock of 21 or feeding alone. From Fig. 4.3, its best option is to join the group, even though this means a decreased fitness for all the other group members. The same is true for the next arriving bird. Indeed the group grows to 54 before the 55th bird does as well feeding alone as joining the group. The same is true for subsequent individuals while the group size stays at 54. Although, if the 55th bird to arrive decided to feed alone, the 56th bird would do best to join it in a group of two, rather than feeding alone or making a group of 55. By this argument, we would not expect to see optimally sized groups in nature. Rather, we would expect to see groups of greater than optimal size. See Box 4.1 for a related arguement.

In the case above, the final group size (which we'll call the Sibly size) is more than twice the optimum. Although this need not always be the case, and the relation between the optimal group size and the Sibly size depends on the shape of the fitness curve. Further, at the Sibly size, group members do only very slightly better than a lone individual. Note that the individual that brings the group up to 21 members experiences a huge increase in fitness (compared to foraging alone), whereas an established group member suffers only a relatively small fitness decrease (compared to being in a group of 20), as a result of another individual joining the group. The newcomer stands to gain more from joining than an existing group member can gain from excluding it. Hence exclusion of the newcomer is unlikely, and group size will increase.

> **Box 4.1** An illustrative example of Sibly's argument that groups will generally be larger than optimal
>
> Sibly (1983) gives another thought experiment. Imagine a situation where there are nine groups, of sizes 18, 19, 19, 20, 20, 20, 21, 21, and 22. The most disadvantaged individual is allowed to move to the group that benefits it most. This continues until no individual benefits from moving. We assume that fitness is symmetric around the optimum of 20 as in Fig. 4.3. The most disadvantaged individuals are in groups of 18 and 22. Let us assume that one of the individuals in the group of 22 moves. It should join the group of 19, forming a group of size 20. Let us assume that an individual from the group of 18 moves next, it should join the remaining group of 19. We now have group sizes of 17, 20, 20, 20, 20, 20, 21, 21, and 21. The most disadvantaged individuals are in the group of 17, the next five individuals to move should all leave this group and join a group of 20. We now have group sizes of 12, 21, 21, 21, 21, 21, 21, 21, and 21. Each of the next eight movers goes from the small group to each of the other groups. We now have a group of 4 and eight groups of 22. Each of the group of 4 then join separate groups, and we end up with four groups of 22 and four groups of 23. Now no individuals benefit from moving. We have moved from nine groups with sizes around the optimum to eight groups, all larger than the optimal size. We recommend that the reader plays this game with pencil and paper a few times using any starting distribution of group sizes they like. You should observe that redistribution generally increases group sizes, and that the fraction of individuals in groups larger than the optimum generally increases.

4.2.2 Refinements of the argument

Sibly's work has prompted considerable theoretical development. Giraldeau and Gillis (1985) demonstrated that the optimal group size can be stable if the fitness function is such that the fitness associated with a group size one larger than the optimal size is less than the fitness associated with being alone (see Fig. 4.4). Kramer (1995) pointed out that Sibly's arguments require the condition that individuals act sequentially. If small subgroups were able to move together, then the group of 54 individuals described above would not develop, since, for example, a group of 40 individuals would split into two groups of 20. Whilst Kramer's contention that 'if small subgroups can break off and join other such subgroups then the population easily reverts to optimal group sizes' seems a little sweeping, we certainly agree that such group movements should result in group sizes much closer to the optimal. If a group of 54 were split into subgroups by some external factor (say in response to a predatory attack), then we would not generally expect them to re-form into a single group again. Kramer also takes issue with Sibly's argument for not allowing individuals to move after they have joined the group. Imagine that individuals can only move singly again, and we have a group of 54 and a group of 2. An individual from the group of 54 would now benefit from switching groups. This procedure continues until we have much more balanced group sizes. Again, this will act to reduce the extent to which group sizes exceed the optimum. However, biologically, there may be circumstances where such movements are not possible. For example, if the groups

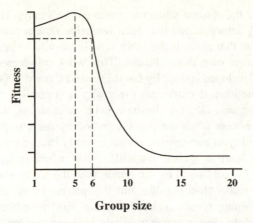

Fig. 4.4 A fitness function for which the optimum group size is stable. This is possible because the fitness of individuals of all larger group sizes (e.g. $5+1=6$) is less than that of lone individuals. (Adapted from Giraldeau and Gillis 1985.)

are colonies of nesting birds, then once an individual has build a nest and laid eggs in it, it may not be willing or able to invest the energy in starting from scratch in another colony, even if that colony would be more attractive to a similar bird that had not already begun investing.

4.2.3 The role of relatedness

Another complicating aspect is relaxation of the assumption that individuals are only interested in maximizing their own fitness. If individuals are genetically related then this may not be true. Now an individual may reject joining a group when the benefit to its own direct fitness is exceeded by a reduction in its inclusive fitness due to the direct fitness of related individuals in the group decreasing. Clearly, this conflict does not occur if the group size before joining is suboptimal, since all individuals would then benefit from the group size increasing. However our prediction would be that such inclusive fitness considerations would tend to curtail increases in group size far beyond the optimal. Increasing relatedness between individuals would tend to lead to a decrease in group size, moving closer to the optimal size.

Another assumption of Sibly's work is that, when an individual decides to join a group, the other group members can do nothing about this. If we assume that the group can prevent the entry of an outsider, then we would predict that optimum group sizes would then be found. However, again the situation could be changed by relatedness. The group might allow the group size to increase beyond the optimal if the potential joiner were related to them, and each group member's loss of direct fitness is more than compensated for in their inclusive fitness by the direct fitness increment given to the joiner. In contrast to the situation directly above, where group entry was controlled by the joiner, we would expect increased relatedness to cause an increase in group size above the optimal when the group controls entry. However, Giraldeau and Caraco (2000) demonstrated that such inclusive fitness considerations cannot result in a group size that exceeds the Sibly size that would be produced by individuals maximizing their own direct fitness. This work (pioneered by Higashi and Yamamura 1993, developed further by Giraldeau and Caraco 2000) leads to empirically testable predictions. If entry into the group is controlled by the joiner, then characteristic group size should decrease with relatedness, however it should increase with relatedness when entry is controlled by the group (in the sense that groups have the ability to bar entry, it seems unlikely that groups can often kidnap an individual and hold it against its will). Also, when group sizes are below optimum, and entry is controlled by the joiner, then the joiner should preferentially choose to join kin rather than non-kin, but this preference should be reversed for situations where joining would make the group size more than the optimal. In contrast, groups should always prefer to accept kin over non-kin.

It is clear that theoretical predictions depend strongly on whether the decision to expand a group is taken solely by the potential joiner or is controlled by those already in the group. This is a complex situation and, as Higashi and Yamamura (1993) pointed out, may change with group size. Even when current group members are related to the potential joiner, there will be times (when group size is at or above the optimum) when entrance would benefit the joiner's inclusive fitness, but not that of the current group members. When such a conflict occurs, it is likely that the situation will be resolved in favour of the side that is prepared to invest more in forcing their way. When the group size is closer to the optimal than to the Sibly size, the potential joiner has much to gain from entry into the group, relative to the cost to current group members. Thus we might expect this to be a circumstance where joiners are able to force their will on groups, because they should be prepared to invest more than current members in any conflict. The same is not true if the group size is closer to the Sibly size, now the potential joiner will only receive a small benefit from gaining entry into the group, and so will be less willing to invest time and energy in trying to achieve this aim. However, the situation will also be influenced by ecological circumstances, and in particular by the way inter-individual conflicts are resolved in that species. It is also important to remember that the costs to the group of barring entry can be shared among a number of individuals.

4.2.4 The influence of competition

Zemel and Lubin (1995) suggested another reason why individuals may be prepared to stay in groups that are larger than optimal. Under some circumstances, an individual striking out on its own would do better, but only if it can get far enough away from its current group that there is no competition between it and them. In circumstances where this is impossible, it may pay to remain in the group. One can imagine this occurring in a group of large carnivores within a limited reserve. We imagine circumstances such that if an individual were moved to another identical reserve, then it would do better than it currently does in the group. However if it leaves the group but remains within the current reserve then it does badly, because it is often out-competed for food by the group that it left behind.

Another important consideration is that when group members are intrinsically different from each other, then they may differ in what they consider to be an optimal group size. We will consider this more fully in Chapter 6.

4.2.5 The effect of dominance hierarchies

Optimal skew models (Vehrencamp 1983a; Keller and Reeve 1994) have been used to explain grouping of breeding females (see Chapter 7). In the basic model, the dominant individual in a group controls the proportion of total reproduction that is allocated to each subdominant. Subdominants each require some proportion in order for group breeding to be a better option for them than breeding alone. If the dominant gains some advantage from being in a group, then it may benefit it to yield some of the total reproduction to subordinates as a 'staying incentive'. It will offer subordinates the minimum amount required to keep them in the group. If yielding this much reproduction is the best alternative for a dominant (compared to becoming subordinate or breeding alone) then a stable group forms. These models predict that the proportion of reproduction yielded to subordinates should be high when relatedness between individuals is low, the benefits of grouping are low, and the competitive abilities of subordinates are only just less than that of dominants.

Hamilton (2000) has extended this theory to consider foraging groups. His model assumes that the ability of a group to gather food is an increasing but saturating function of group size. There is a single dominant that controls the share of foraging resources given to subordinate joiners. Current joiners cannot influence subsequent joiners, and the dominant can only do so by withholding resources. This leads to a stable group size that is larger than the optimal group size under equal division of resources between all group members (the optimal group size discussed earlier), but smaller than the group size where equally sharing individuals do only very slightly better than solitary individuals (the Sibly size). Both the dominant and the subdominants agree on this intermediate, stable group size.

In an illuminating variation, dominants can only control a portion of the resources gathered, the rest being shared equally among all group members. If this portion is small, then in some cases a subordinate benefits from joining even when the dominant offers none of the resources that it controls. Under such circumstances, the group can grow to a larger size than is optimal for the dominant. Group size is now controlled by subdominant joiners, who now do not agree with the dominant on optimal group size. A logical extension to this work would be to allow joiners to choose between groups.

4.2.6 Empirical evidence for active recruitment to foraging groups

Informing others of newly discovered food sources is common among eusocial animals (Judd and Sherman 1996, and references therein). The benefit to the informer in these cases can easily be understood in terms of increasing their inclusive fitness.

However there have been several examples of apparently active recruitment in non-eusocial vertebrates reported in the literature. In such cases, the benefits that the signaller may obtain are less obvious. These reports are reviewed here with the aim of demonstrating that, whilst several different mechanisms have been proposed to explain the evolution and maintenance of such signals (often termed food calls) in terms of fitness benefits to the signaller, these mechanisms have generally gone untested.

The first paper to provide experimental evidence that individual foragers actively recruited others was Elgar (1986). This paper reported house sparrows (small passerine birds) feeding on either a whole piece of bread, an equivalent amount of bread broken into crumbs, or birdseed. For each observation, one of these food types was placed on an artificial feeder on a rooftop. The roof was surrounded by a low wall. If the food was a whole piece of bread, then the first sparrow to arrive would not vocalize but would begin to feed alone. For the other two food types, the first arrival would alight on the wall and emit 'chirrup' calls before feeding. Elgar (1986) demonstrated that these calls attract other sparrows to share the food. His conclusion was that sparrows called when the food was divisible because competition costs were lower than for a single piece of bread that can only be handled by one bird at a time, and because the caller gained some benefit from the presence of others. This benefit was postulated to be reduced risk of predation or time spent in anti-predator vigilance; previous experiments, Elgar *et al.* (1984), were cited as support.

Playback experiments reported in Elgar (1986) demonstrated convincingly that 'chirrup' calls attract conspecifics. First arriving birds modify their calling in response to the type of food available (Fig. 4.5a), they are more likely to forage after they are joined (Fig. 4.5b), and individuals in larger groups 'chirrup' less (Fig. 4.5c) — all strongly suggesting that calling is used facultatively by the birds. The benefit to the calling individual that has led to the evolution of this behaviour is less clear. A lack of sexual differences in calling causes Elgar to dismiss explanations based on mate attraction. Facilitation in finding food could be dismissed as the food had already been discovered by the first arrival. Nor were numbers required to gain access to the food, which was freely available and could be accessed by a single individual. In support of the theory that sparrows gain anti-predatory benefits from grouping, Elgar *et al.* (1984) demonstrated that individual vigilance rates decrease with increasing flock size. However, whether this change was mirrored by an increase in feeding rate was not recorded. Nor was the experiment designed to quantify the effect of group size on predation rate. Thus although Elgar's hypothesis that first arrivals' benefit from summoning others through the much discussed anti-predatory advantages of grouping is plausible, this hypothesis was not rigorously tested.

Similar to the 'chirrup' call in house sparrows, a squeak call has been detected in cliff swallows that is associated only with food (swarms of flying insects), and which attracts conspecifics (Brown *et al.* 1991). The hypothesized benefit that callers gain is explained as follows:

"The dense insect swarms on which these birds feed often are ephemeral. An individual probably cannot exploit any given concentration for very long and thus incurs little cost in sharing it with conspecifics. Alerting other birds increases the number of foragers in the vicinity of the

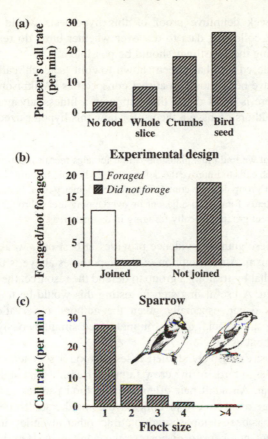

Fig. 4.5 (a) The first house sparrow to arrive calls little if no food is available, more if there is food but as one large piece of bread, more still when the bread is broken into crumbs, and most when bird seed is available. Quoted values are means across birds. (b) When crumbs were the food source, sparrows were more likely to feed if joined by another and more likely not to feed if not joined. (c) The mean call rate of an individual group member as a function of group size. (Adapted from Elgar 1986.)

swarm, increasing the odds that the insects' subsequent movements will be tracked by a least some members of the group. Even if other birds do not also call, the caller could benefit via local enhancement simply by watching the nearby group members as some of them track the subsequent movements of the prey. Calling to inform other foragers would be especially useful when birds are feeding nestlings and thus must commute to and from the colony and foraging grounds. Informed foragers would track the swarm's movement while the caller goes back to its nest."

We find this mechanism very plausible, although there is no data in the paper that can be used to critically test it. In defence of the authors, we admit that designing experiments to seek definitive proof of this hypothesis would be challenging. However, at least collecting data to discover whether birds do return to the same swarm after feeding their nestlings should be possible.

Greater spear-nosed bats have been shown to emit 'screech' calls when feeding, and these calls have been shown to attract conspecifics (Wilkinson and Boughman 1998). Again, there is no evidence that bats gain a fitness advantage from attracting others. The authors do have a theory for how this hypothesized advantage may come about.

"The hypothesis that we find most plausible is that females recruit roost-mates into foraging groups using screech calls to improve the defence of predictable feeding sites. The most compelling evidence for group defence comes from radio-telemetry data. Females from the same social group repeatedly foraged in adjacent or overlapping areas throughout the year, with females from different groups typically foraging in distinct areas."

In our view, these authors' evidence provides at best only weak support for the theorized mechanism. An assumption of the hypotheses above is that there is food competition and that by forming a group to defend the resource, the number of competitors is reduced. A useful first step in testing this would be to explore whether competition does occur, especially given the authors' observation that 'captive females from the same social group routinely feed simultaneously from the same piece of fruit without aggression'.

One of the most extensively studied cases of food-associated calling is that by juvenile ravens (large birds of the crow family). Here, at first sight, the benefit to callers seems clear. An adult pair often defends food patches (animal carcasses). Only when in large numbers can juveniles overcome the adults' defence. However, although calling has been demonstrated to attract other juveniles, it does not appear to attract sufficient numbers to dispossess the adult pair (Heinrich and Marzluff 1991; Heinrich *et al.* 1993). Rather, the researchers conclude that 'the large numbers of ravens assembled at carcasses can only be explained by recruitment from a nocturnal roost' in addition to any attracted by calling *in situ*. They suggest that the importance of initial calling is to recruit individuals that may have access to a roost that the initial caller does not. However evidence for this is lacking, and so it may be that calling serves only to reduce aggression by adults and other juveniles (as discussed by these authors).

Hauser (2001) observed food calls in field experiments on rhesus monkeys. On finding food, monkeys sometimes emitted calls. These calls were associated with increased numbers of other monkeys joining the first individual at the food source. Individuals that found food but did not call, ate significantly less food from artificial food sites than monkeys that did call. Hauser suggests that this is linked to non-callers experiencing greater aggression from joiners than callers do. In effect, non-callers are punished for their lack of co-operation, if their 'crime' is detected. We find Hauser's experiment and conclusions to be convincing. Calling benefits because non-callers are punished. This explains how calling can persist once such a punishment system is established, but leaves open the question of how it evolved.

Caine *et al.* (1995) studied captive red-bellied tamarins (small New World monkeys) and found calls that functioned to attract others to a food source. They could find no proximate benefit to the caller, since aggression at the food source was common. They conclude that their 'data was consistent with (but not a direct test of) the notion that food calling may ultimately benefit the caller by keeping group mates near'. Tamarins live in small groups that move rapidly as they forage; an individual that stops to feed may run the risk of losing touch with the group, but the likelihood and consequences of this are untested. Similarly, the suggestion of Chapman and Lefebvre (1990) that spider monkeys use calling to gather large groups in order to gain vigilance or other anti-predatory benefits is entirely speculative.

Janik (2000) found that bottlenose dolphins made specific calls that were associated with encountering food and that lead to an approach by conspecifics. However, Janik argued that the benefit of anti-predator protection commonly used to explain a functional benefit for callers seems unlikely to apply to this system, given that adult dolphins have very few predators. He also discards arguments based on avoidance of aggression by other members of the social group, because 'fish are very mobile and it is very unlikely that other dolphins will find the same food spot at the same time'. Other possible explanations are inclusive fitness benefits through attracting kin, increasing social status, and increasing feeding success through co-ordinated attacks on shoals of fish. However he emphasizes that there is as yet no evidence for these. Furthermore, Janik suggests that the calls are possibly not directed at other dolphins, as they are concentrated at frequencies to which dolphins are not particularly receptive. Hence, his favoured hypotheses is that callers gain a fitness benefit from calling by stunning or modifying the behaviour of fish, and that other dolphins take advantage of this, possibly at a cost to the caller. No experimental data are available to support this, although there is increasing evidence that cetaceans can use sonic energy to stun fish.

Careful study of so-called food calling in junglefowl (an Asian gamebird) has shown that such vocalizations are much more strongly associated with courtship than with feeding (van Kampen 1994, 1997; van Kampen and Hogan 2000). Similarly, the interesting suggestion that male domestic chickens can increase copulations by calling to attract females to discovered food (Marler *et al.* 1986) has yet to be tested.

There is no doubt that a wide range of taxa show behaviour whereby signals that are produced by an individual that discovered food are used by others to locate the food source. A great deal has been written about the mechanisms by which the individual that discovers food can benefit from signalling. However, before we allow these theories to become embedded in behavioural ecology textbooks, we need to critically test them. As we have demonstrated, empirical support is currently very much lacking. We hope that the preceding discussion will motivate critical testing and will provide some pointers as to how this might be best achieved. Particular attention should be given to demonstrating the following four points:

1. The signal (believed to be a food call) is not obligate, in that it would be possible for the signaller to forage as effectively without sending the signal.

2. The signal is associated with food.
3. The sender receives a fitness benefit from signalling.
4. This benefit outweighs any costs associated with its use.

4.3 Observed group sizes in nature

4.3.1 Social carnivores

Empirical testing of optimal group size ideas has focussed very heavily on hunting groups of social carnivores: principally lions, wild dogs, and wolves. The classical study is by Caraco and Wolf (1975), who analysed the data on lion foraging from Schaller (1972). They predicted an optimal foraging group size of two, but found that observed groups were significantly larger than this. This discrepancy has been explained in terms of kin selection arguments (Rodman 1981) or the idea of stable group sizes being larger than the optimum, as discussed earlier in this chapter (Giraldeau and Gillis 1988). Clark (1987) developed a model assuming that lions were not seeking to maximize food intake rate, but to minimize the risk of failing to meet some maintenance requirement: the predictions of this were more in line with the observed group sizes in Schaller's data. Packer *et al.* (1990) presented further data on lions. They found that when prey were abundant, neither the mean nor the variance in per capita daily intake rate was affected by group size. In fact, in times of scarcity, singletons, or animals in groups larger than five did better (by both measures) than those in groups of two, three, or four. However, during times of scarcity, lions were found to use groups of two, three, or four much more often than the theory predicted. Packer *et al.* (1990) suggested that this discrepancy was due to the demands of co-operative cub defence and/or group territoriality making hunting in very large or very small groups less attractive. They call for more research on lions in a different locality, a call that was answered by Stander (1992). This study found singletons to be particularly disfavoured in times of scarcity and small groups (under four) to be favoured in times of plenty. However, hunting alone was always uncommon, but hunting in large groups was more common in times of plenty than energy maximizing would predict. In summary, we are still some way from agreement on the factors driving foraging group size in lions.

The situation is similar for wild dogs. Creel and Creel (1995) presented a very comprehensive data set. After some further developments (Packer and Caro 1997; Creel 1997), it seems clear that no matter the currency considered, hunting groups seem to be considerably smaller than optimal. This is difficult to explain, especially as large group size allows better defence of kills against hyenas (Fanshawe and FitzGibbon 1993; Carbone *et al.* 1997). The problem may be related to difficulties in identifying the appropriate optimization currency for coursers like wild dogs, where the energetic expense of hunting may be considerable.

Schmidt and Mech (1997) analysed all the available data for wolves, and came to the rather surprising conclusion that per capita food acquisition rate declined with increasing pack size. They then suggested that wolf packs should be seen as a form of extended parental care. Packs primarily consist of an adult pair and their offspring. The offspring remain with the parents to acquire hunting experience and subsequently disperse to breed themselves. Thus grouping increases offspring survival but does not *per se* increase hunting efficiency. The effect of group size on various fitness surrogates has also been investigated for several other group-hunting taxa, e.g. humans (Smith 1981), chimpanzees (Boesch 1994), killer whales (Baird and Dill 1995), but these species are if anything harder to study than those discussed above, and detailed data is very much lacking.

4.3.2 Data from other taxa

Although group-hunting vertebrates are very charismatic, their long lifespan, complex social ecology, and low population density make their foraging behaviour difficult to study in the field, and impossible in the laboratory. Perhaps we have been guilty of concentrating too much on these species, especially since the number of mammalian predatory species known to engage in group hunting is small, estimated at 22 by Boesch (1994). The attractions of invertebrates can be seen clearly in the study of colonial spiders by Avilés and Tufiño (1998). Colonies can be kept in the laboratory, and colony size can be manipulated so that a full curve of fitness against group size (like Fig. 4.3) can be constructed from replicated experiments. The simpler ecology and short lifespan of this species means that the effect of group size on all the major components of fitness can be measured, as can reproductive fitness itself. Such laboratory experiments predict that a group size of around 50 should be optimal. However, naturally occurring colonies tend to be much larger or smaller than this. The authors suggest that the likely source of the difference is that the theory's assumption that individuals are free to leave or join a group at any time does not hold for social spiders. Colony growth occurs by recruitment from within the colony, with some individuals leaving to attempt to form new colonies (often unsuccessfully) after reaching a certain life history stage. This work, combined with our wealth of understanding of social spider ecology and their amenability to manipulation and measurement, would suggest that such taxa might be a more appropriate test-bed for (possibly further developed) theory than the charismatic mega-fauna of the African plains.

4.3.3 Distribution of group sizes

When we look at the distribution of animals between groups within a species, we generally find that they show a right-skewed frequency distribution, where relatively small group sizes are common but a long tail of larger groups appears to the right (see Fig. 4.6a). However, it is very instructive to re-plot such data in terms of the group size experienced by each individual. This can show that most individuals experience intermediate group sizes, with relatively small fractions of the population being in small or large groups (Fig. 4.6b). There has been an extensive theory developed

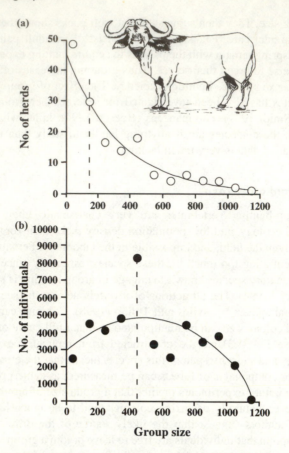

Fig. 4.6 The distribution of group sizes from a study of African buffalo reported by Sinclair (1977). (a) Numbers of different herd sizes observed. This illustrates that small herds are common, with medium and larger herd sizes being progressively less common. (b) The same data re-plotted from an individual's perspective: the number of individuals that are in herds of different sizes. This illustrates that most individuals are in herds of intermediate size, with relatively few individuals in large or small herds.

to explain the observed distribution of group sizes within populations (Bonabeau *et al.* 1999; Flierl *et al.* 1999, and references therein). These models generally seek the simplest set of rules for the fusion and splitting of groups that generates group size distributions like those observed in nature. However, the search for simple rules tends to lead to characterization of individuals as passive 'victims' of mixing processes. Such an approach ignores the ample evidence that natural selection will act on traits influencing the group sizes experienced by an individual. That is, it ignores the idea that individuals should have a preference between group sizes. There is strong evidence that such group size preferences exist (Pritchard *et al.* 2001). Hence, we suggest that the time is right for a change in emphasis in the simulation modelling of distributions of group sizes. We need to move away from seeking the simplest models that agree with observed group size distributions towards the simplest models that agree with empirical data on group size distributions *and* with the observed behaviour of individuals. Such models would allow us to explore the evolution of such behaviours, as well as providing models that should have more predictive power to describe how the expected frequency distribution of group sizes is likely to be affected by ecological factors. The over-simplified behavioural rules of current models do not at present allow these interesting avenues to be explored; see Chapter 9 for a fuller discussion of this issue.

4.4 Summary and conclusions

In theory, testing whether observed groups are of an optimal size or are larger is quite straightforward. One only needs to experimentally reduce the naturally observed group size slightly. If the original group size was optimal, then the manipulation will lead to a reduction in the individual fitness of remaining group members. If, however, the original group was larger than optimal, then the manipulation may well lead to a fitness increase. Although this type of experiment is conceptually simple, it has not proved as popular as unmanipulated observation of hunting groups in large carnivores. As Section 4.3.1 outlines, the second approach has not provided definitive answers to whether or not animals are found in group sizes that are above optimal. We believe that the time is ripe for manipulative experiments on species (such as colonial spiders) that are more amenable to this approach than large mammalian carnivores.

The recent theory on control of group entry and relatedness discussed in Section 4.2.3 provides clear predictions that can be empirically tested. For example, if entry into the group is controlled by the joiner, then group size should decrease with relatedness, however it should increase with relatedness when entry is controlled by the group. Also, when group sizes are below optimum, and entry is controlled by the joiner, then the joiner should preferentially choose to join kin rather than non-kin, but this preference should be reversed for situations where joining would make the group size larger than the optimal. In contrast, groups should always prefer to accept kin to non-kin. Empirical testing of the assumptions and predictions of these recent models would seem a key step in advancing our understanding of grouping in nature. In particular, theoretical predictions depend strongly on whether the decision to expand a group is taken solely by the potential joiner or is controlled by those already in the group. Understanding when and how existing group members influence further additions to the group is an essential next step.

As we discuss in Section 4.2.6, a wide range of taxa show behaviour whereby signals emitted by a discoverer of food are used by others to locate the food source. A great deal has been written about the mechanisms by which the discoverer can benefit from signalling, but empirical support is currently very much lacking. We hope that our discussion will motivate critical testing.

There is a need for further theoretical work also. An important extension to previous work would be to allow lone individuals a choice between groups of different sizes, rather than a choice of joining a given group or being alone. Similarly, allowing existing members of a group the opportunity to join other groups, as an alternative to striking out on their own, would be an important step towards greater ecological realism. The work of Hamilton (2000) seems a particularly useful foundation for such developments.

5

Spatial heterogeneity of costs and benefits within groups

5.1 Introduction

Chapters 2 and 3, on the benefits and costs of grouping, assumed that the fitness consequences of grouping apply more or less equally to all individuals in a group (and are only a function of group size). This simplification will be challenged in this chapter, in which we explore the importance of spatial positioning within a group for individual fitness. Animal groups vary greatly in their spatial structure (Parrish and Hamner 1997) and there is growing evidence that the costs and benefits of group-living are strongly related to the spatial position of an individual within a group. Fitness is notoriously difficult to measure and so, in this chapter, we will use a number of surrogates to estimate the fitness consequences associated with different spatial positions in groups: these include energy intake (Section 5.4.1) and expenditure (Section 5.4.2), risk of predation (Section 5.5) and parasitism (Section 5.6), and access to mates (Section 5.7).

This chapter reviews the fitness differences that accrue from different spatial positions in a group. The mechanisms underlying positioning behaviour (i.e. the different forms of inter-individual distance regulation by group members), however, will be deferred to Chapter 9 on self-organization. In the first part of this chapter (Section 5.2), we discuss the problem of defining spatial positions within groups. Some discrepancies and/or misunderstandings in the literature are due to the fact that spatial positions were not clearly defined. Clear definitions are essential to making studies reproducible and comparable. Another common problem (Section 5.3) with studies of grouping animals is that cause–effect relationships are often not clearly identified. If we observe behavioural differences between individuals in different spatial positions of a group, we need to be sure whether these differences were caused by group geometry or by other factors (such as inter-individual differences in internal state). In Sections 5.4.1 and 5.4.2 we discuss the influence of group positions on energy intake and expenditure, which is followed by a consideration of the net-energy payoff in Section 5.4.3. The other major factor we consider with regards to group position is predation (Section 5.5). First, we review the empirical support for Hamilton's theory of marginal predation concerning stationary groups (Section 5.5.1). Then we discuss how Hamilton's model can be extended to the case of mobile groups (Section 5.5.2) pointing out the dearth of empirical tests of existing theory in the area. Much of the theory that is reviewed in the section on predation can also be applied to parasitism (Section 5.6) as far as ectoparasites that take a single blood-meal from their host are concerned. In Section 5.7, we look at the influence of spatial positions on reproductive success, discussing some of the literature on colony-breeding and lekking. In Section 5.8 we investigate the effects of social hierarchies on positioning behaviour. The underlying assumptions are that individuals in groups partly compete for the greatest share of the benefits of grouping but that they also differ in their strategy of prioritizing different factors (due to intrinsic inter-individual differences and variation in internal states). This takes us to the final topic of trade-offs and how animals evaluate group positions when several different and potentially conflicting factors are taken into account (Section 5.9).

5.2 Group structure and spatial positions: definitions

To investigate how different spatial group positions vary with regards to the above fitness currencies we have to take a closer look at the spatial structure of groups. Furthermore, we have to make assumptions about the ways in which grouped animals search for prey and are attacked by predators. In an influential paper, Hamilton (1971) showed that if a predator could appear at any spatial position, and always attacked the nearest prey, then the attack risk of an individual can be calculated using Voronoi polygons which quantify the domain of danger for a given group position (Fig. 5.1) (see Okabe *et al.* 1992 for the calculation of Voronoi polygons). An alternative and more general way of quantifying the predation risk of group positions is to run a computer simulation in which reiterated predator–prey

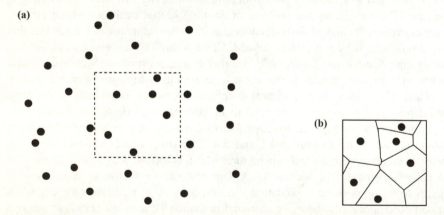

Fig. 5.1 (a) Top view of a two-dimensional group of animals whose position is symbolized by filled circles. (b) For a subset of animals the area size of the Voronoi polygon is calculated which can be used to quantify the predation risk of individual group members.

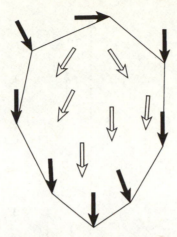

Fig. 5.2 Top view of a two-dimensional group where an individual (symbolized as an arrow) is defined as peripheral (filled arrow) if it is at the vertex of the smallest convex polygon which encloses the entire group.

encounters are staged for given group geometries and predator attack modes. The frequency with which different positions are attacked by a predator can then be taken as an indication of relative risk (Bumann *et al.* 1997).

A third option requires finding suitable operational definitions for delineating different functional groups of spatial positions. Animal groups come in many different shapes and sizes and the challenge is therefore to develop definitions of group positions that are generally applicable but also unambiguous. In most cases, these definitions concern front and back positions and the edge and centre ones because these are areas that are most likely to experience fitness differences. Krause and Tegeder (1994) suggested defining peripheral individuals as those that are at the vertex of the smallest closed convex polygon that includes all group members (Fig. 5.2). The definition of front and back positions seems obvious as long as we have a direction of movement: an individual is considered to be in the front half of a group if it is in position 1 to $n/2$ (numbering from the front, where n is the membership size) for shoals of even size and in position 1 to $(n-1)/2$ for shoals of odd size. However, this definition does not take details about the shoal geometry into account and becomes problematic when several individuals are aligned. One way of dealing with the latter issue is to calculate the centroid of the group and assign front and back positions relative to it (Fig. 5.3).

5.3 Cause and effect relationships

One of the problems with collecting data on the fitness consequences of group positions is that the cause–effect relationship of observed events is not always as

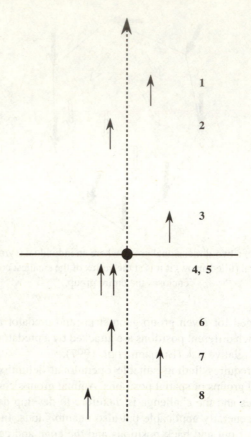

Fig. 5.3 Individuals (symbolized as small arrows) are numbered according to their position in the group relative to the group's direction of locomotion that is indicated by a stippled arrow. The centroid is marked by a black spot and the line bisecting the centroid at a right angle to the direction of the group can be used to define front (positions 1–3) and back (positions 4–8) of the group.

obvious as it seems. For example, individuals at the edge of a group may feed at a higher rate because:

(a) The food availability is higher on the edge than in the centre.
(b) Because they are hungrier (than central individuals).
(c) They are intrinsically more efficient foragers.

Furthermore, individuals in the centre and on the edge may differ in their perceived predation risk resulting in different vigilance rates, which then might affect the time available for foraging. These mechanisms are not mutually exclusive, and careful experimental manipulation is usually required to single out the main cause (or causes) for an observed difference in feeding rates. Most studies (including the ones discussed below) provide information on only one or two of the above factors when measuring feeding rates and there is a need for further, more systematic experimental studies.

5.4 Energy gains and losses

5.4.1 Energy intake

There are numerous studies indicating that both food availability and food quality are higher for individuals in peripheral and front positions of groups compared to centre and back ones (Table 5.1). The study by DeBlois and Rose (1996) is particularly noteworthy because of the large size of the fish school considered, over 10 km long. They found that front fish had significantly fuller stomachs that contained higher quality food than rear fish. Similar differences in foraging success between

Table 5.1a Evidence for higher food availability and prey capture rates in peripheral positions of groups (adapted from Krause 1994b)

Species	Food availability edge > centre	Prey capture rates edge > centre	Source
European starling *Sturnus vulgaris*	Yes	/	Keys and Dugatkin 1990
Barnacle goose *Branta leucopis*	Yes	Yes	Black *et al.* 1992
Colonial spider *Metepeira incrassata*	Yes	Yes	Rayor and Uetz 1990
White ibis *Eudocimus albus*	Yes	/	Petit and Bildstein 1987
Mussel *Mytilus edulis*	/	Yes	Okamura 1986
Wood-pigeon *Columba palumbus*	/	No	Murton *et al.* 1966, 1971

Table 5.1b Evidence for higher food availability and prey capture rates in front positions of groups (adapted from Krause 1994b)

Species	Food availability front > back	Prey capture rates front > back	Source
Roach, chub *Rutilus rutilus, Leucicus cephalus*	Yes	Yes	Krause *et al.* 1992, 1998b; Krause 1993b
Atlantic cod *Gadus morhua*	Yes	Yes	DeBlois and Rose 1996
Northern anchovy *Engraulis mordax*	Yes	Yes	O'Connell 1972

fish in the front and back were found in small shoals (consisting of a few dozen fish) of cyprinid fish (Krause 1993b), suggesting that these trends are largely independent of group size.

If some positions are better for foraging than others, we might expect to see preferences for such positions particularly when individuals are hungry and/or have an increased energy demand. This has indeed been demonstrated in the roach, a cyprinid fish, and whirligig beetles, in which individuals that had been experimentally food-deprived preferred positions in the front and periphery of the group whereas well-fed ones were found more in the centre and back (Krause 1993b; Romey 1995). Similarly, parasitized minnows and killifish (whose energetic requirements tend to be higher due to the parasite burden) were located more on the group edge than the centre (Krause and Godin 1994a; Barber and Huntingford 1996; Ward *et al.* 2002).

5.4.2 Energy expenditure

Studies on energy expenditure in different group positions are mostly concerned with the costs of locomotion and the loss of heat to the environment. Theoretical studies on the costs of locomotion in fish shoals and in bird flocks showed that there is considerable potential for energy savings for individuals in the back of a group (Weihs 1973, 1975; Lissaman and Schollenberger 1970; Norberg 1989) (Fig. 5.4).

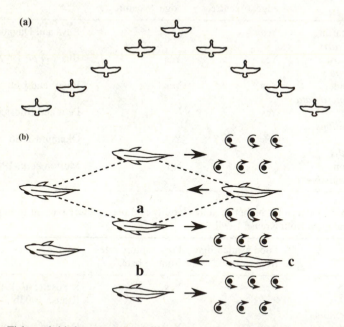

Fig. 5.4 Fish and birds can potentially save energy by adopting certain formations: (a) V-shape in birds (adapted from Heppner 1997) and (b) diamond pattern in fish. Fish **c** benefits from the flow (straight arrows) induced by the vortices produced by fish **a** and **b**. (Adapted from Weihs 1973.)

However, empirical demonstrations of differences in energy expenditure between group positions are few and, in the case of fish shoals, also controversial (Partridge and Pitcher 1979; Abrahams and Colgan 1985; Pitcher and Parrish 1993; Herskin and Steffensen 1998). Observations of reduced tail beat frequencies in connection with a decrease in oxygen consumption rate in groups of seabass probably provide some of the best empirical evidence to date of individuals having to work less in rear positions in fish shoals (Herskin and Steffensen 1998).

Fish (1995) explored the energy saving associated with formation swimming in ducklings, and found that the 'end' duck benefits the most. He hypothesized a positional trade-off for ducklings, between the energy saving of being at the end of the line and the protection of being close to the adult at the front. He argued that rigid bodies gain the most effective saving by being in line astern, but this is not true for laterally oscillating bodies such as swimming fish. This idea could easily be explored using Bill and Herrnkind's (1976) method for estimating the energy of movement of lines of spiny lobsters (see Chapter 2, Section 2.8.2). One could simply test the relative effort required to pull different arrangements of the same individuals along at the same speed, to test whether single file is the optimal arrangement. Even for a single file, exploration of the effects of changing inter-individual distance would also be illuminating. Bill and Herrnkind report no obvious relation between size of individuals and place in naturally occurring lobster queues; again a simple extension of their experimental protocol would allow exploration of the consequences of different arrangements of a collection of different sized individuals.

In many shapes of formation fliers, the energy savings are quite different for birds in different positions (Hummel 1995). This heterogeneity of benefit is especially true of groups of heterogeneously sized birds. Birds with larger wing spans produce greater energy saving for others, an effect which is felt most strongly by the bird directly behind (Hummel 1995). Some game-theoretical modelling on this subject could yield testable predictions about the formation, dynamics, and structure of flocks.

Cluster-roosting has been frequently observed in various species of birds: long-tailed tit (Smith 1978), European wren (Armstrong and Whitehouse 1977), and bats (Kurta and Fujita 1988). It seems intuitively obvious that differential energy savings from centre and peripheral positions will accrue in these cases but little systematic study has been done.

5.4.3 Net-energy payoff

Very few detailed studies are available that have taken both the energy intake and expenditure into account to calculate estimates of net-energy gains or losses for different group positions. An exemplary study is the one by Black *et al.* (1992), who found that in grazing barnacle geese energy expenditure in central and peripheral positions was approximately the same whereas energy intake was higher in peripheral birds resulting in a 27% higher net-energy intake for individuals on the edge.

5.5 Predation risk

The predation risk of different group positions can be measured indirectly by recording the frequency of attacks (directed at a group position) or as a per capita mortality rate. The former has the advantage of being more commonly observed, but suffers from the problem that the attack-to-capture ratio could be different for different group positions, in which case attack rates would not be a reliable indicator of risk. However, frequent attacks, even if they do not lead to increased mortality, can still be associated with a substantial cost to the prey, in terms of disrupting foraging behaviour and mating activities. Another way of obtaining indirect information on the predation risk of group positions is to selectively frighten a single individual in a group (Krause 1993c) whose subsequent positioning behaviour should be an indicator of its perception of predation risk of different group positions. Actual prey captures are the most valuable type of information but they are relatively rarely observed in the field, which makes quantitative analysis difficult in some cases. In the following sections, we will focus on laboratory and field studies that reported predator–prey encounters resulting in actual prey mortalities, since they are likely to provide the most relevant information on position-related risks.

5.5.1 Stationary groups

Hamilton (1971) stated that, if predators always attack their nearest prey, then we should expect peripheral individuals to be at a higher risk than central group members. Although there has been some controversy over whether predators always attack the nearest prey (Parrish 1989), Hamilton's prediction of higher risks on the edge has generally found good empirical support (Table 5.2a). If peripheral positions are more dangerous, then we might expect this to manifest itself in the anti-predator behaviour of some species. A fascinating study by Rattenborg *et al*. (1999) revealed that ducks in peripheral group positions keep the eye open that faces away from the group centre but close the other. Presentation of a predation stimulus to the open eye of a peripheral bird resulted in a rapid response by the individual. Birds are capable of unihemispheric sleep which means that one half of the brain will stay awake (controlling the open eye) while the other one is asleep (connected to the closed eye). This is a phenomenon otherwise only known from aquatic mammals, where it serves the function of making swimming possible during sleep. In birds, however, a functional interpretation was lacking until recently. Rattenborg *et al*.'s (1999) work suggests that the ability to control sleep and wakefulness simultaneously in different areas of the brain serves an anti-predatory function.

5.5.2 Mobile groups

In contrast to the good empirical support for Hamilton's model regarding stationary groups, much less information is available on differential predation risks in mobile

Table 5.2 Evidence for position-related mortality in (a) stationary (edge versus centre) and (b) mobile groups (front versus back); only those studies were included that observed actual prey mortality and not just attack rates (adapted from Krause 1994b)

Species	Predation risk	Source
(a)		
Colonial spider	Edge > centre	Rayor and Uetz 1990, 1993
Metepeira incrassata		
Uganda kob	Edge > centre	Balmford and Turyaho 1992
Kobus kobthomasi		
Pacific damselfish	Edge > centre	Foster 1989
Abudefdu saxatilis		
Water flea	Edge > centre	Jakobsen and Johnsen 1988
Bosmina longispina		
Mussel	Edge > centre	Okumura 1986
Mytilus edulis		
Bluegill sunfish	Edge > centre	Dominey 1981;
Lepomis macrochirus		Gross and MacMillan 1981
Sandwich tern	Edge > centre	Veen 1977
Sterna s. sanvicensis		
Fieldfare	Edge > centre	Andersson and Wiklund 1978
Turdus pilaris		
Brewer's blackbird	Edge > centre	Horn 1968
Euphagus cyanocephalus		
Blackheaded gull	Edge > centre	Kruuk 1964;
Larus ridibundus		Patterson 1965
Adelie penguin	Edge > centre	Taylor 1962
Pygoscelis adeliae		
(b)		
Creek chub	Front > back	Krause *et al.* 1998a
Semotilus atromaculatus		

animal groups (Table 5.2b). Field observations of predation are rare and often cannot be clearly related to the spatial position within the group. Furthermore, accurate assessments of the number of individuals in peripheral and central positions are hard to obtain for groups that are in motion, which makes the calculation of per capita risks difficult (Lindström 1989; Parrish *et al.* 1989). Not surprisingly therefore, most studies on this topic were conducted in the laboratory, where the behaviour of both predators and prey can be filmed and subsequently analysed. Jakobsen and Johnsen (1988) investigated fish predation on swarms of Daphnia (small freshwater crustaceans) and found that individuals in peripheral positions incurred a much higher risk than central ones. However, the extent to which a swarm of Daphnia is capable of directional locomotion is debatable (Table 5.2). In contrast, Parrish (1989) found that central Atlantic silversides (a small marine fish species) experienced a higher per capita attack rate from black seabass than peripheral conspecifics. However, only five prey captures were observed in over a thousand attacks, which either means that seabass are not a very efficient predator of silversides, or that the conditions under which the experiments were run were not conducive to seabass making captures. In either case, we cannot draw any firm conclusions regarding position-related predation risks.

Hamilton's model has mainly been discussed in the context of differences between central and peripheral positions. In mobile groups, however, we might expect to find an additional risk gradient from the front to the back (of a group), if we use the same simplifying assumption that predators attack the nearest prey (Bumann *et al.* 1997). An additional level of complexity can be introduced by predators being either stationary or mobile. If the predator is stationary, then front individuals receive almost all of the attacks and back ones are virtually never affected. If the predator is mobile, then individuals in back positions get attacked occasionally but still not as often as front individuals (Fig. 5.5).

There is some support for front positions to be more risky (than other positions) from a study on fish shoals (Krause *et al.* 1998a) but more empirical work is needed on attack strategies by predators. In particular it is unclear whether some predator species have preferences for attacking certain areas of a group. It has been frequently suggested that predators' strikes are aimed at splitting up the group to isolate individuals that can then be caught more easily (Major 1978; Parrish 1993). The latter has been observed in predators of pelagic fish shoals such as tuna (Parrish 1993) and in raptors attacking large flocks of passerine birds. In these cases, an initial strike by the predator is usually followed by a chase, and it remains unclear whether this type of attack behaviour puts some group positions at a higher risk than others. Clearly if the initial strike is mainly aimed at splitting the group (and less at prey capture), then the focus should be on that stage of the attack when the predator successfully removes an individual from a subgroup and on that individual's spatial position.

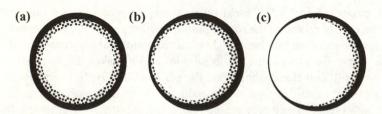

Fig. 5.5 Predation risk is indicated by the concentration of black dots showing the position of attacked group members for a group of 100 individuals that are randomly distributed within a circular area defining the group shape. (a) Group and predator are stationary. (b) Both group and predator move continuously with equal velocity. (c) Group moves and predator is stationary. A total of 10 000 predation events were simulated for each of the three models. (Adapted from Bumann *et al.* 1997.)

5.6 Parasites

In Section 2.3, we showed that some of the mechanisms (e.g. encounter–dilution and Selfish herd effects) that lower per capita predation in groups can also be beneficial to group members in reducing risk from parasites. Therefore it may not be surprising that we find similar positional effects. Peripheral individuals in herds of reindeer incurred significantly higher parasite loads than central ones (Helle and Aspi 1983). And in domestic cattle it has been observed that the individuals walking at the front of the group have higher tick loads than other group members (Newson *et al.* 1973). Marking has shown that dominant individuals were more likely to be found in front positions and there was also a certain degree of inter-individual variation in tick susceptibility that was due to immune response (or physical reactions such as rubbing off of ticks) unrelated to the spatial position in the group (Newson *et al.* 1973), which emphasizes the need for carefully controlled studies.

Groups of hosts should be more conspicuous to their parasites than single individuals and thus attract larger number of parasites. If parasite avoidance is the main function of group formation (Mooring and Hart 1992), then we should not expect the edge individuals to have higher parasite loads than single individuals. Otherwise the group would disperse. If, on the other hand, parasite avoidance is just one of several benefits gained through grouping, then edge individuals may well carry a higher parasite load than singletons because higher parasite loads might be traded-off against other benefits. The above discussion applies mainly to blood-sucking flies and ticks. Little information on position-related costs and benefits is available to date on parasites and diseases whose transmission is aided by close spatial contact between individuals.

5.7 Reproductive success

In colony-breeding and lekking species, reproductive success is generally higher for individuals in central positions. The former is a result of a number of factors. Egg and/or offspring mortality tend to be higher in peripheral positions and greater exposure to the physical environment and movement of conspecifics through the colony result in a higher probability of damage to the nests (Kruuk 1964; Dexheimer and Southern 1974). Such positions will also benefit less from communal defence. Again studies must be careful to control for confounding factors such as the breeding colony tending to be centred on the most suitable habitat type for nesting, then expanding out into less preferred habitat types.

In lekking species, males that occupy central positions are generally preferred by females, which could be due to central positions being an indicator of mate quality and/or to a reduced predator risk in central positions (Gosling and Petrie 1990; Fryxell 1987; Balmford 1991).

5.8 Dominance status

In Sections 5.4–5.7 we have reviewed a number of fitness-related factors that vary with group position (foraging efficiency, risk of predation, and parasitism). Dominance is not one of them because the dominance status does not change depending on position but conversely, positioning potentially depends on status. Therefore dominance can be considered as a factor that potentially allows individuals to obtain a greater share of the grouping benefits by monopolizing spatial positions. Barta *et al.* (1997) predicted that dominant individuals should be found more in the centre of groups, whereas subordinate ones should be more in the periphery. This prediction was based on the idea that dominant ones could easily monitor subordinate group members from the centre, and the centre being a good position from which to scrounge food discovered by group mates (Fig. 5.6). There is evidence from a number of studies supporting the idea that dominant individuals are found more centrally in groups, and that this behaviour is often context-dependent. In several ungulate species, dominant individuals were observed to take up central positions when fleeing from predators (Hirth and McCullough 1977). Larger fish

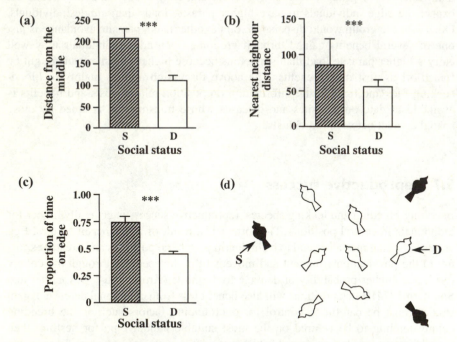

Fig. 5.6 Differential positioning predicted for dominant (D) and subordinate (S) birds in a flock. Dominant birds which scrounge from other flock members are predicted to (a) occupy flock positions closer to the centre, (b) have smaller nearest neighbour distances, and (c) spend less time on the edge of the flock than subordinates. (d) Positioning of dominant and subordinate flock members. (Adapted from Barta *et al.* 1997.)

occupied central positions in shoals of chub, a small freshwater cyprinid, when under predation threat (Krause 1994a). Dominant spice finches were observed in central positions more often than on the periphery in foraging flocks (Flynn and Giraldeau 2001), and dominant primates have been reported to occupy central positions in fruit trees (Janson 1985, 1990). Evidence of aggressive interactions being responsible for differential positioning was found in colonial spiders, where larger females (which carry egg sacs) replaced smaller ones from the centre, presumably because the centre provides a safer rearing environment for the offspring (Rayor and Uetz 1990). However, careful observations are needed to show that differences in positioning behaviour between group members are due to competitive interactions (where all individuals prefer the same positions but only the dominant ones can get them) and not simply a by-product of differences in trade-offs between dominant and subordinate group members. Using the example of colonial spiders again, peripheral positions may be more suitable for younger spiders because they allow for fast growth even though they are more risky. Whereas older females with egg sacs potentially give priority to reproduction over high foraging rates. The latter issue leads directly into the debate on trade-offs that is discussed in the following section.

5.9 Trade-offs between different fitness currencies

When a predator attacks a group, the most important (if not the only important) issue for all group members is the minimization of predation risk. If certain group positions are safer than others, then we should expect to see competition for these positions within the group, with the dominant group members being most likely to take up the safest spots. However, these moments of direct threat from a predator tend to be relatively brief and infrequent in an animal's life. For the majority of the time, the choice of a position should be influenced by multiple factors such as maximization of foraging efficiency and minimization of predation risk (to name but two of the most important ones). From the information above, it should be clear that those positions that are particularly good for foraging generally also have the highest predation risk, which requires trade-offs to be made. The problem of combining different factors such as foraging efficiency and predation risk, which are measured in different ways—(one being quantified as food intake and the other as a risk)—into a single fitness-related currency has been tackled using stochastic dynamic programming (Mangel and Clark 1988; Houston *et al.* 1988; see Krebs and Kacelnik 1991 for a review). However, few studies have attempted to apply this approach to the positioning behaviour of individuals within groups.

Some of the best information on the dynamics of positioning comes from colonial spiders (*Metepeira incrassata*) (Rayor and Uetz 1990, 1993) that live in colonies of 250–1500 individuals. Prey availability is considerably higher in peripheral positions but so is the predation risk. Removal experiments established that larger females with egg sacs had a strong preference for central positions, presumably because of the great importance of the survival of offspring. For younger spiders, however, peripheral positions may actually be advantageous because they have yet to attain the larger body size necessary for successful reproduction. If true, this would indicate a shift of trade-offs from being risk-prone to being risk-adverse resulting in differential positioning with age and size.

In some species of shoaling fish, where position changes can be accomplished very quickly, we might expect to see a dynamic rotation of shoal positions. Hungry individuals might prefer front and edge positions, where high foraging rates can be achieved at the price of high risks, with fish switching to safer central positions with increasing stomach fullness.

Frequent positional shifts have been reported from birds in formation flight (Gould and Heppner 1974). Although to what extent turbulence drives these changes remains unexplored.

5.10 Summary and conclusions

A large part of the work on positional differences is descriptive and careful experimental manipulation is needed to distinguish between alternative explanations. Clear definitions of group positions are important in this context to make empirical work on different study systems comparable and reproducible.

Little conclusive work has been done on mobile groups, such as bird flocks and fish shoals. This is mainly due to problems with recording the behaviour of entire groups, which often take up considerable space, and with the great speed of events, which requires high speed video analysis techniques that have only recently become available (and affordable). In particular, studies concerning medium- and long-term positioning behaviour of individuals are largely missing. Improved tagging and marking techniques for individual identification, in combination with new software that allows the simultaneous tracking of multiple mobile objects, should create many possibilities for experimental work and push the boundaries in this field of research.

Predation events are rarely observed in nature, and not easy to stage in captivity without loss of realism. Computer simulations of predator–prey encounters are one possible way in which predictions for real-life encounters can be made. As more information about the hunting strategies of predators and anti-predator behaviour of prey becomes available, it should be possible to extend the work by Hamilton (1971) and Bumann *et al.* (1997) to create sophisticated individual-based simulations in which group members can change their spatial position during an attack in order to reduce risk.

Another challenge is to develop models in which the costs and benefits of group positions are expressed in a common currency so that predictions for short- and long-term positioning behaviour of individuals resulting from trade-offs can be made. Stochastic dynamic modelling is a promising technique for approaching this problem, and has already been applied to similar problems in other contexts (Clark and Mangel 2000).

6

Heterogeneity and homogeneity of group membership

6.1 Introduction

Until now, this book has discussed groups as if they were composed of intrinsically identical individuals. Of course, this is often a poor approximation of reality. Even in cases where there is marked similarity between individuals in a group, we must consider if individuals have chosen to assort non-randomly. The next sections look at assortment in the context of both anti-predatory behaviour and foraging in groups. However, other factors, such as transport costs or disease transmission, although much less studied, are also likely to have a considerable influence on the composition of groups.

As Section 6.2 demonstrates, we are quite far from a theory that allows predictions regarding the phenotypic assortment in groups that we should expect to see in nature. The number of papers describing models that can be used to explore non-random assortment of ·individuals can be counted on the fingers of one hand. However, these papers do form suitable foundations for development. We suggest how such developments might best arise.

Section 6.3 considers one mechanism by which assorted grouping may come about: non-random prey choice by predators. For example, predators may preferentially target the least common phenotype in a group, in order to combat the confusion effect. Such selectivity should produce evolutionary pressure for grouping with similar individuals, and may even lead to mimicry, as discussed in Box 6.1. Experimental evidence of this selectivity (often called the oddity effect) is limited. Further, we argue that some experimental results on this are much more widely quoted than others, which has probably lead to a premature belief in the pervasiveness of the oddity effect in nature. There is also empirical support for instances where predators show a preference for the most common phenotype and where preference is independent of the relative abundance of different prey types. In comparison to the oddity effect, predicting the effects of these types of predation pressures on group structure is less clear, although we do present a theoretical structure for considering this.

Section 6.4 is devoted to fish, with which a great deal of empirical work has been devoted to exploring shoaling preferences, with respect to factors such as species, size, colour, parasite load, relatedness, familiarity, and competitive ability.

Section 6.5 argues that there may sometimes be benefits to being in a mixed species foraging group rather than associating only with your own kind. This association can be either parasitic or mutualistic. Mutualistic interactions can occur when species have complementary abilities or if they have different resource gathering needs, such that the benefits of grouping can be experienced without paying a cost in competition.

Optimal group size and optimal group composition will often vary for intrinsically different individuals. As Sections 6.6 and 6.7 consider, there has been a dearth of study on how between-individual conflict over group size is resolved. It is likely that many individuals will spend a significant proportion of their time in suboptimal groups, but how individuals seek to minimize the negative consequences of this is still very much unexplored territory.

6.2 Theory of assortativeness

Very unusually for behavioural ecology, theoretical investigations into assortment are much less common than empirical ones. Until recently, the only significant theoretical work was that of Ranta *et al.* (1994), developing ideas from Lindström and Ranta (1993). This simulation model assumed that the ability of a group of fish to find food was an increasing function of both group size and the body lengths of individual group members, larger fish being more effective in food finding. Once a food patch had been discovered, then the reward was shared among the group members, weighted by their length, with larger fish getting proportionately more. Simulations were performed, with individuals being allowed to change group (or forage on their own) if this led to an improvement in their net food reward rate. This model predicted assortment when the effect of individual length on foraging success was strong. This suggests that a small fish does not do well in a group of larger individuals because, although the group as a whole finds many food patches, the small individual's return from those patches is very small. The small fish would do better by getting a greater share of a smaller number of patches. Conversely, large fish do better with other large fish, because for them the increase in patch discovery rate compensates for the reduced share from each patch, compared with the situation where the big fish is with smaller individuals. More theoretical work is needed to explore the generality of these results to changes in parameter values and model assumptions. These workers also report that assortment is affected by the total number of fish in the environment, with increasing numbers of fish reducing the tendency for assortment. This is a particularly interesting result, as it invites empirical testing.

Conradt and Roper (2001) present a model that seeks to explain assortativeness through what they term 'activity synchrony'. They suggest that many of the advantages of being in a group only occur if animals perform a given activity (e.g. foraging or resting) synchronously. They suggest that when animals in a group differ (say in species, sex, or size), then they pay greater costs for achieving synchrony, because they have differing priorities for one activity over another. Conradt and Roper further suggest that this should make groups of differing individuals less stable than groups of similar individuals. They develop a simple model of random fusion of individuals into groups and increased group fission for groups of differing individuals, and use this to explore the importance of activity synchrony in explaining sex-assorted grouping in red deer. While some of the details of this model (such as animals showing no preference for same-sex animals as group mates) are open to challenge, this seems another exciting avenue for further theoretical work.

6.3 The influence of predation on assortment

A predator that attacks group-living or aggregated prey is often faced with the decision as to which prey individual to choose, unless it can consume the entire group (as may occur in some cases, such as whales feeding on entire krill swarms). This means that a predator has a simultaneous choice between several prey items (as opposed to sequential prey encounters that have been discussed elsewhere: Beukema 1968; Gendron 1986). Our objective is to investigate how predators select prey under these circumstances, what the underlying mechanisms for their choice behaviour are, and how this affects prey group composition (through the evolution of particular grouping strategies in prey populations).

Empirical studies have shown that predator preferences for particular prey individuals are often phenotype-dependent (Landeau and Terborgh 1986) (see Box 6.1).

Box 6.1   The influences of differential predation on group structure

Differential predation means that different phenotypes (within a prey group) are not selected in proportion to their numerical abundance. Phenotype-dependent preferences can take different forms:

(a) Predators can have an inherent or acquired preference for a particular phenotype over another, regardless of the relative frequencies of phenotypes (because it is easily captured or of high energy content, for instance).

(b) Predators can show a frequency-dependent preference, i.e. they have a preference for rare or common phenotypes.

In practice, we would expect that often both types of preference act simultaneously.

Consequences of frequency-independent predation

Let us first consider the case in which predators have no frequency-dependent preference for prey phenotypes. We assume a scenario in which larger individuals are preferred over small ones by predators, because of their higher nutritional value. If faced with a choice between two groups of the same size, an individual (no matter whether it is large or small) should join the group with the larger proportion of large individuals because its predation risk in the event of an attack would be lower. However, such a preference can only evolve if the benefits of a reduced risk in the event of an attack are not outweighed by the costs of increased attacks directed at the group (i.e. the high proportion of larger individuals may

Box 6.1 (*Cont.*)

make the group more conspicuous or more attractive to predators). Furthermore, we might expect to see trade-offs between group size and group composition. An individual might be better off joining a small group of large individuals than joining a large group of small individuals.

Consequences of frequency-dependent predation

The consequences of frequency-dependent predation for prey groups depend on whether we deal with a system in which predators prefer the rare or the common prey phenotype. The common phenotype should only be preferred by predators that are unconstrained by the confusion effect. This type of preference should lead to the evolution of polymorphic prey groups because any prey phenotype that dominates the group numerically will be selected against. If, however, the rare phenotype is preferred, prey groups are selected for phenotypic homogeneity. Any individual whose phenotype stands out from the rest will incur a much higher risk than other group members.

The preference of predators for phenotypically odd prey has another interesting aspect: it could potentially lead to the evolution of mimicry. A less abundant species *A* could gain a benefit from grouping with a highly abundant species *B* in terms of being better protected against predators in larger groups. However, for this benefit to be fully realized, individuals of species *A* would have to resemble individuals of species *B*. Otherwise they would appear as phenotypically odd and the advantage of joining a larger group might be outweighed by the oddity effect. In populations where a certain overlap of phenotypic characteristics between different species already exists, a preference of predators for odd prey could thus become the driving force behind the evolution of mimicry rings that could involve a large number of different species (Ehrlich and Ehrlich 1973). The group size effect would also benefit the mimicked species *B*, greatly increasing the stability of such mimicry rings. A first example of this form of mimicry has been described by Dafni and Diamant (1984) who found a system in which an otherwise solitary fish species mimics a shoaling species to gain protection from grouping with this species.

6.3.1 Predator preference for odd prey (the oddity effect)

The suggestion that unusual looking (or otherwise odd) individuals in a group might be preferentially attacked, or more easily captured, by predators has a long history. The idea is intuitively appealing, the argument being that by concentrating on an individual that looks characteristically different from the rest, a predator can counteract the confusion effect described in Section 2.4. Early quantitative studies of this effect failed to consider that odd individuals in a group may also be intrinsically more conspicuous against the background (rather than against their group mates), or that predators may have an inherent preference for such individuals, regardless of their frequency within a group. The first study to critically test these alternative hypotheses was by Ohguchi (1978) using different coloured water fleas as prey and sticklebacks (small fish) as predators. This study was able to demonstrate an increased vulnerability of odd-coloured individuals in a group that could not be explained by either of the alternative explanations given earlier in this paragraph.

The study of Landeau and Terborgh (1986), using normal silver and blue-dyed prey fish (minnows) attacked by a larger predatory fish (bass), also provided evidence for the oddity effect. Monocoloured groups of sizes one, two, and four suffered high predation rates, whereas bass were almost always unsuccessful in capturing individuals from monocoloured groups of eight or 15 fish. However, when one or two of the individuals in a shoal of eight were a different colour from the rest, then the predator's success rate was greatly increased (Fig. 6.1). Most, but not all, of these captures were of odd individuals. This suggests that odd individuals were particularly vulnerable to attack, but also that other individuals suffered a greater predation risk through the presence of these odd individuals. These results are often cited; however, some other results in this paper have not received the exposure that they deserve. Odd individuals still appeared to benefit from joining groups compared with being alone. Lone fish were captured successfully in every trial, whereas an odd individual in a group of eight became the victim on only 50% of trials. Further, an individual in a group of eight made up of four of each colour was less vulnerable than a monocolour group of four and only marginally more vulnerable than an individual in a monocolour group of eight (Fig. 6.2). Having one or two odd individuals in a group of 15 did not increase the predator's ability to make captures. The authors conclude

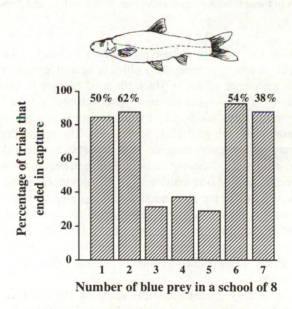

Fig. 6.1 Percentage of 5 minute trials that ended with prey capture, as a function of the composition of a school of eight fish. The percentage of occasions in which it was an individual of the minority phenotype that was captured is given above some of the bars. (Redrawn from Landeau and Terborgh 1986.)

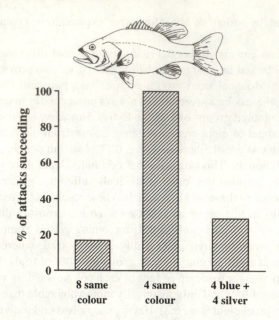

Fig. 6.2 Comparison of the percentage of attacks that were successful against a group of eight same-colour fish, a group of four same-coloured fish, and a group of eight composed of four of each of the two colour types. (Redrawn from Landeau and Terborgh 1986.)

that 'the effect on the majority phenotype in being joined by one or more individuals of another phenotype is generally positive or neutral, except under a relatively narrow range of parameter values in which the total number of group members is small and the ratio of phenotypes is strongly imbalanced'. These very interesting results deserve much more widespread exposure, and would greatly benefit from empirical exploration of their generality. Some studies do suggest that the generality of this result may be less than complete. Peuhkuri (1999) found that fish given a choice between two groups, did not change their preference for joining a group depending on the number of odd individuals in that group. However, in these trials, preference was also unaffected by introducing visual contact with a live predator. This may suggest that shoal choice was being made for reasons other than protection from predation. Similarly, the observation of Wolf (1985), that fish left a group more readily in the face of a simulated predation threat when they had few conspecifics in the group seems counter to one's expectation from the work of Landeau and Terborgh (1986).

The only other widely quoted empirical study on the oddity effect is that of Theodorakis (1989), which explored oddity in body length for minnows attacked by bass. In groups of 25 large and five small fish and in groups of 25 small and five large, the minority size was taken more often than would be predicted by chance.

6.3.2 Evidence for the oddity effect from prey behaviour

Peuhkuri (1997, 1998) used foraging activity as an index of predation risk, in a group of mixed small and large sticklebacks. She found that the feeding activity of small individuals was insensitive to group size. However, larger fish foraged less when in a group of predominantly small ones. Indeed, their foraging rate increased as their relative frequency in the group increased. This could be seen as consistent with the oddity effect inducing the large fish to be more vigilant when different from other group members. However, somewhat against this interpretation, this effect was independent of the presence of a predator or of group size. It seems likely that these results are confounded by changing competition for food (as well as changes in individual predation risks) as group composition changes. Allan and Pitcher (1986) found that three fish species formed mixed species shoals under controlled conditions, but sorted into individual species on the addition of predator cues. In contrast, Krause (1994a) found that the addition of an alarm substance did not change group composition within mixed-length shoals of chub (cyprinid fish), but caused reorganization of groups such that neighbours were more likely to be of similar size, larger (and probably socially dominant) fish being near the centre and smaller fish on the periphery (see Chapter 5).

6.3.3 Predator preference for common prey

Preference for the more common prey phenotype is often referred to as 'apostatic selection' (Endler 1991). One of the best documented examples comes from chickens feeding on pellets of two different colours, brown and green (Fullick and Greenwood 1979). Chickens showed an inherent preference for brown pellets but there was also

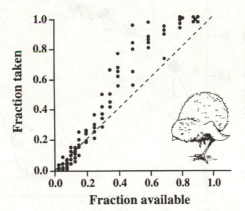

Fig. 6.3 The fraction of brown pellets taken by a chick in relation to the fraction of brown pellets offered to it in a mixed presentation of brown and green pellets. The points should fall on the line if the chick was unselective and had an equal preference for each pellet colour. However, the data indicate a preference for the more common type. (Redrawn from Fullick and Greenwood 1979.)

a strong frequency-dependent preference for the more common pellet type (Fig. 6.3). Since the pellets were stationary, the predator was probably unconstrained by the confusion effect (which led to a preference for odd prey in Section 6.3.1). In the absence of such a perceptual constraint, the birds could maximize their intake rate by developing a so-called 'search image' for brown pellets (Endler 1991). It is known that some predators, such as sticklebacks, can switch between both types of preferences (for the more common or the rare prey type) depending on what the situation requires (FitzGerald and Wooton 1993).

6.3.4 Frequency-independent preferences

Differential predation is not always dependent on the frequency of the prey phenotypes. Predators often choose their prey based on inherent or acquired preferences due to differences in prey profitability or palatability, or due to the prey's ability to flee quickly or to defend itself. Murdoch and Stewart-Oaten (1975) offered predatory shore snails differing proportions of two species of mussel (*Mytilus californianus* and *M. edulis*). When the proportions were equal, snails showed a strong preference for the thinner-shelled *edulis*. The line in Fig. 6.4 represents the predicted consumption of *edulis* if the snail's preference remained the same, but the relative proportions of the two mussels in the environment changed: this provides a good description of the observed predatory behaviour of the snails.

Pike (a predatory fish) prefer slender prey fish in mixed shoals of deep-bodied and slender crucian carp (of the same body length), probably because they are easier to

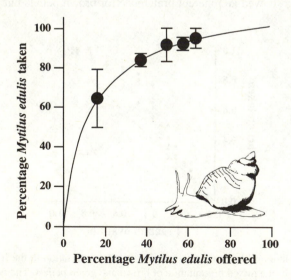

Fig. 6.4 The fraction of one type of mussel eaten by a snail in relation to the fraction of that type offered to it in a mixed population of two mussel species (mean ± S.E.). The curve predicts a constant preference for one prey type that is unmodified by relative availabilities. (Redrawn from Murdoch and Oaten 1975.)

swallow. An interesting aspect to this is that crucian carp can change from a slender to a deep-bodied form in response to chemical cues from increased local pike activity, and thereby attain better protection from predators (Brönmark and Miner 1992; Brönmark *et al.* 1999).

The concept of frequency-independent differential predation can be very useful in trying to explain the evolution of multi-species groups. Wildebeest, for instance, are a preferred prey of many large African carnivores. Therefore the association of other ungulate species (such as zebras) with wildebeest could be advantageous for the zebras in terms of reduced predation risk at no cost for the wildebeest (Sinclair 1985). Imagine a highly idealized situation where prey (wildebeest and zebra) are static and lions appear at random in the environment and attack an individual within a fixed radius of its position. Let us also assume that given a choice of targets, the lion will select a wildebeest rather than a zebra. If zebra position themselves close to wildebeest, then this reduces their vulnerability, since on most occasions when they are in the circle for possible attack, so will a wildebeest. However, the presence of the zebra has no effect on the wildebeest, which would have exactly the same chance of being targeted if there were no zebra. Indeed, the wildebeest may also gain protection from an increased group size, if the presence of other animals causes, for example, a confusion effect, making attacks less likely to succeed. FitzGibbon (1990), who studied predation by cheetah on Grant's and Thomson's gazelles, invoked a similar argument to explain mixed species grouping in these two species. Both species gain from being in larger groups, whether mixed or single species. However, there were added benefits to being in a mixed species herd for the more vulnerable Thomson's gazelle. Cheetah showed a preference for attacking single species Thomson's gazelle herds

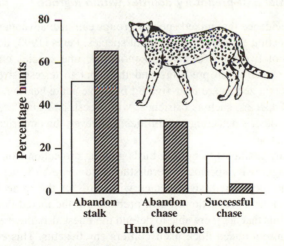

Fig. 6.5 The effect of group composition on the outcome of 209 cheetah hunts on gazelles. Mixed species groups (hatched bars) consisted of both Grant's gazelles and Thomson's gazelles and single species groups (open bars) consisted only of Thomson's gazelles. (Redrawn from FitzGibbon 1990.)

compared to mixed species ones (Fig. 6.5), and were more successful against single species ones (both results controlling for herd size). However, on attacking a mixed species group, the cheetah was more likely than chance to chase a Thomson's gazelle. This may provide added protection to Grant's gazelle. Thus both species may prefer multi-species herds to similar sized single species ones.

6.3.5 Reduced vigilance in mixed species groups

It has been suggested that mixed species avian flocks can have all the advantages of group vigilance with reduced competition costs (compared with a single species group), if the species involved exploit different resource niches (Sasvari 1992). This idea gains support from studies by Metcalfe (1989) on reduced aggression within mixed species flocks of shore-birds. However, Metcalfe (1984) points out that the vigilance benefits may not be as high as first supposed. It may be that different species have some, but not all, predators in common. In his study, larger species of birds did not respond to imminent attacks by predators that were not a threat to them, but were to other smaller species within the flock. The generality of these results in unclear. Pius and Leberg (1997) did not find reduced aggression rates in mixed species sea-bird colonies. Kristiansen *et al.* (2000) demonstrated that white-fronted geese reduced vigilance when there was a pre-breeding pair of greylag geese (which maintain high vigilance rates) in their flock, compared with the same site without the greylags. Barnard *et al.* (1982) also studied shore-birds, and suggested that plovers actively joined lapwings, despite lapwings stealing food from them, because of the anti-predatory vigilance benefits.

6.3.6 Differential anti-predatory abilities within a group

There is some evidence that mixed species groups can take advantage of the different predator detection abilities of constituent species. Peres (1993) describes how in a mixed group of two types of tamarin monkeys, one species characteristically reacted first to shared avian predators and the other to terrestrial ones. Diamond (1981) suggests that zebras are near-sighted but have acute hearing. By associating with far-sighted species, such as ostriches, wildebeest, and giraffes, they can each complement the other's detecting capabilities. However, the empirical basis for this suggestion is unclear.

There are many examples of nesting birds gaining protection from close association with more aggressive species. Several studies compare birds nesting in a mixed colony alongside more aggressive species with birds of the same less aggressive species nesting alone. A reduction in nest predation in the mixed flock might come about because birds that are more able to defend their nest also prefer mixed colonies, rather than because a mixed flock itself confers any benefits. This explanation was critically tested and rejected for sparrows nesting with gulls by Wheelwright *et al.* (1997). Another explanation, that the mixed species colonies tended to be in sites with intrinsically lower predation, was not supported by the study of Post and Seals

(1993). Hence, it does look as though less aggressive species do actually benefit from proximity to more aggressive ones, at least in birds. A particularly interesting example is that of bar-tailed godwits seeking protection by nesting with long-tailed skuas. Larsen (2000) described how rodents are the skuas' main prey, but in years of rodent scarcity the eggs and chicks of other birds are taken. Godwits showed a greater propensity to nest with skuas in years of rodent abundance. Bogliani *et al.* (1999) discussed the propensity of woodpigeons to nest near hobbies (small birds of prey). Although adult woodpigeons were sometimes attacked by hobbies, proximity to hobby nests greatly reduced woodpigeon egg and nestling predation by crows. More interestingly yet, woodpigeons were found to preferentially nest near those hobbies that show more vigorous nest defence behaviour. It is also worth considering that there may sometimes be a cost to the protecting species. Groom (1992), in another avian study, reported that nighthawks obtain greater fledging success by nesting near terns that aggressively mob predators. However, terns nesting with large numbers of nighthawks spend more time in defence and have reduced fledging success.

6.4 Evidence for the evolution of group mate preferences in prey

The arguments of the previous sections predict that, when the oddity effect operates, there should be a tendency for individuals to form groups with others of similar appearance. There are several studies on grouping patterns in the wild that are consistent with this prediction (Krause *et al.* 1996a; Peuhkuri *et al.* 1997; Svensson *et al.* 2000). However, these grouping patterns could also be a result of passive sorting mechanisms based on differential locomotion speeds (see Watkins *et al.* 1992 on antarctic krill; Gueron *et al.* 1995 on African ungulates) and/or habitat preferences. To obtain evidence for the evolution of active choice behaviour of group mates, preference experiments need to be carried out. By a significant margin, the great bulk of studies done on group mate choice of individuals has been conducted on fish, which is partly due to the fact that preference experiments are particularly easy to set up for small fish species.

6.4.1 The role of species

A preference for conspecifics over heterospecifics has been reported in a number of fish species including rock bass (Brown and Colgan 1986), female Trinidadian guppies (Magurran *et al.* 1994), banded killifish (Krause and Godin 1994b), Atlantic salmon, and rainbow trout (Brown *et al.* 1993). Keenleyside (1955) pointed out an important feature of species association tests. Whether or not a significant test result in favour of a preference for conspecifics is obtained may largely depend on which species is offered as an alternative. This should in most cases be another shoaling and sympatric species, in order to provide a biologically relevant test. Keenleyside

(1955) observed a preference for conspecifics in threespine sticklebacks, when the heterospecific stimulus fish were bitterlings, but not when roach were used. The functional significance of preferring to group with conspecifics is likely to be due to two main factors. By associating with conspecifics an individual reduces its chances of suffering the increased predation risk of the 'oddity effect'. A similar case can be made for foraging behaviour. The probability of detecting suitable food is likely to be maximized in the company of conspecifics that have similar dietary preferences. However, the degree to which the latter is counter-balanced by post-detection competition is arguable and needs further testing.

6.4.2 The role of body length and colour

Association preferences for size-matched fish have been found in numerous studies (reviewed in Ranta *et al.* 1994). This can occur both when shoal-mates are conspecific and when they are heterospecific (Krause and Godin 1994b; Crook 1999). Killifish, for instance, have a preference for conspecifics over heterospecifics and for size-matched fish over smaller or larger ones. However, when given a choice between size-matched heterospecifics and larger conspecifics, they strongly preferred the heterospecifics, illustrating how preferences can have different priorities when conflicts arise (Krause and Godin 1994b). Size assortment has been often found to commence (Krause 1994a) or increase (Ranta *et al.* 1992) in response to a perceived predation threat. Fish have been found to assort by size within a single shoal (e.g. Pitcher *et al.* 1986; Ward and Krause 2001, both European minnows; Krause 1994a, chub), or split into several size-assorted shoals (Ranta and Lindström 1990, threespine stickleback).

Some studies have also investigated preferences for stimulus fish based on body patterns and coloration. McCann *et al.* (1971), using black and white photographs as the stimulus, showed that zebrafish preferentially associated with fish with the normal stripe pattern. McRobert and Bradner (1998) reported that domestic mollies of two-colour morphs, which had been kept together in separate tanks, preferentially associated with fish of matching colour.

If individual differences in competitive ability correlate with certain phenotypic characters, theory predicts that shoals should become assorted by that phenotype (Lindström and Ranta 1993; Ranta *et al.* 1993). Given that small fish are often poorer competitors than their larger conspecifics (Ranta and Lindström 1990; Krause 1994a), the preference of small fish for size-assortative shoaling might be explained by avoidance of larger and better competitors, not just by the increased predation risk of appearing odd in a group. Most likely, however, these two selection pressures act simultaneously (Ranta *et al.* 1994), though their relative importance may differ depending on the prevailing ecological conditions.

A number of studies have found free-ranging shoals to be structured by species and size, with small and large fish being found in separate shoals that are often multi-specific but usually numerically dominated by one species (Krause *et al.* 1996b;

Peuhkuri *et al.* 1997). Most of the evidence for size sorting comes from studies on freshwater fish (Hoare *et al.* 2000) but there is some indication that this is also the case for some marine species (Sakakura and Tsukamoto 1996). The above field observations are consistent with laboratory studies that report a strong preference for size-matched individuals and a weaker one for conspecifics. The fact that species preferences can be overridden by body size preferences in the laboratory may be responsible for the fact that multi-species shoals are common and may be even more frequent than single species ones (Krause *et al.* 1996b; Hoare *et al.* 2000).

A correlation between the size of a shoal and the number of species found in it has been observed, but could easily be a statistical artefact (Krause *et al.* 1998c). The larger a shoal, the greater the probability that additional species are found in it by chance. A positive correlation has also been found between the number of species within a shoal and the body size variation, a result that may be promising in terms of providing insights into the mechanisms of shoal formation. There is evidence that both within-species and between-species variation in body size increase with increasing number of species (Krause *et al.* 1998c). The between-species component can probably be explained by the fact that same-sized individuals from different fish species often have different swimming speeds, and that swimming speed generally increases with size. Given that swimming speeds are body size related (Videler 1993) it is not surprising that slow swimming species have to be slightly larger than fast swimming ones in order for both to coexist in the same shoal (Krause *et al.* 1998c). The increase in within-species variation, however, remains unexplained and warrants further attention.

6.4.3 The role of parasitism

Avoidance of conspecifics carrying endo- or ectoparasites has been observed in threespine sticklebacks and banded killifish (Dugatkin *et al.* 1994; Barber *et al.* 1998; Krause and Godin 1996b). There is evidence that parasite detection is based at least partly on visual cues in banded killifish (Krause and Godin 1996b). A prediction based on these preferences for the composition of free-ranging shoals is made difficult, however, by the fact that parasitized fish were also found to prefer unparasitized shoal-mates over parasitized ones (Krause and Godin 1996b). Therefore the composition of free-ranging shoals may depend on potential constraints on the above preferences, such as swimming capacity, which is often reduced in parasitized fish (Videler 1993).

Segregation behaviour in the case of ectoparasites, e.g. *Argulus canadiensis* (Dugatkin *et al.* 1994) potentially reduces the probability of infection. However, fish with endoparasites, e.g. *Schistocephalus solidus* (Barber *et al.* 1998), *Crassiphiala bulboglossa* (Krause and Godin 1996b), which are not directly transmittable between group members, are probably avoided because parasitized shoal-mates may attract predators or may generally be of low quality in terms of shared anti-predator benefits such as predator detection (Krause and Godin 1994a).

6.4.4 The role of familiarity

A robust test of the importance of familiarity in fish shoaling decisions involves generating familiar groups at random in a laboratory environment. Preference for familiars using this method has been demonstrated in the bluegill sunfish (Dugatkin and Wilson 1992), female Trinidadian guppies (Magurran *et al.* 1994; Griffiths and Magurran 1997a,b; Lachlan *et al.* 1998), mixed-sex guppy fry (Warburton and Lees 1996; Griffiths and Magurran 1999), threespine stickleback (Barber and Ruxton 2000), and European minnows (Barber and Wright 2001). In female guppies this preference gradually increases over a period of 12 days (Griffiths and Magurran 1997a), suggesting that a certain degree of temporal shoal cohesion is required before preferences for familiars can become a factor mediating shoal choice.

Associating with familiars may have a number of advantages. Familiarity among the members of a shoal may reduce the fitness costs of competition by reducing aggression between the contestants. Höjesjö *et al.* (1998) found a decrease in antagonistic behaviour with an increase in familiarity among sea trout. Chivers *et al.* (1995) reported that shoals of fathead minnows that originated from the same shoal exhibited greater shoal cohesion, displaying more anti-predatory 'dashing' behaviour and less freezing behaviour, under predator threat, than groups composed of individuals taken from different shoals. It remains to be shown, however, whether this effect is actually due to familiarity between individuals and whether predation risk for the shoal members is lowered as a result.

Fish performing specific behavioural tasks have been shown to choose associates based upon previous experience of the behaviour of these fish. Remaining in a temporally stable group would provide long-term experience of shoal-mates and hence facilitate improved partner choice decisions. In stable shoals of familiar individuals, fish may, for example, use the acquired knowledge about the competitive ability of others and choose the company of those individuals with whom they have been most successfully foraging in the past (Dugatkin and Wilson 1992). Metcalfe and Thomson (1995) demonstrated that fish could discriminate between good and bad competitors, even in the absence of cues such as feeding rate, size, or aggressiveness, and preferred to shoal with poorer competitors. The formation of familiarity-based assemblages is also likely to be beneficial in co-operative interactions, as individuals may then preferentially join individuals that have proven to be most co-operative in the past, e.g. in risky anti-predator behaviours (Milinski *et al.* 1990; Dugatkin and Alfieri 1991).

6.4.5 The role of kinship

Investigation of kin discrimination in fish has mainly focussed on salmonids (see Brown and Brown 1996 for a review). Preference for olfactory cues from sibs rather than non-sibs has been demonstrated in Arctic charr (Olsén 1989), coho salmon (Quinn and Busack 1985), and Atlantic salmon and rainbow trout (Brown and Brown 1992). As all test fish were reared in kin groups, these results may be due to preference for familiar odours rather than an innate kin recognition mechanism.

Quinn and Hara (1986) showed that coho salmon reared with sibs and non-sibs showed no discrimination between them. Furthermore, Winberg and Olsén (1992) showed that young Arctic charr reared in isolation did not show a preference for siblings over non-siblings. A recent study by Olsén *et al.* (1998) however, demonstrated that association preferences of Arctic charr are at least partly based on MHC genotype, suggesting that assessment of the genetic relatedness of individuals may indeed be playing a part in association decisions.

Kin preferences have also been investigated in sticklebacks. Van Havre and FitzGerald (1988) found that stickleback fry preferred to associate with sibs rather than non-sibs when given visual and olfactory cues. This preference was shown by individuals reared in isolation and those reared with non-sibs only, suggesting that sib recognition is innate. More recently, Steck *et al.* (1999) found no preference for sibs in stickleback fry presented with olfactory cues alone.

Fish could gain advantages by choosing to shoal with kin in addition to the direct fitness benefits of shoaling (Pitcher and Parrish 1993). For instance, associating with relatives may increase an individual's fitness because kin are likely to be more co-operative when engaging in risky behaviours such as predator inspection (Milinski 1987). Kin grouping has also recently been demonstrated in the field, most convincingly by Pouyard *et al.* (1999) using mouth breeding tilapia.

Kin preferences have also been reported from frog tadpoles (Rautio *et al.* 1991; Waldman 1984). The reason for this has not been identified, but may be related to selection pressure for alerting kin to predation risk.

6.5 Multi-species foraging groups

Dolby and Grubb (1998, 1999) have been performing manipulative experiments on mixed species flocks of bark foraging birds that form during winter in eastern deciduous forests of North America. These flocks are characterized by nuclear species (tufted titmice and/or chickadees) and several satellite species (downy woodpeckers and nuthatches). The satellite species are socially dominant and can increase their foraging success by kleptoparasitism of nuclear species or by local enhancement. This leads to a foraging cost for nuclear species. All species seem to benefit by being able to reduce anti-predatory vigilance when in the company of others. This may explain why nuclear species tolerate satellites, or it may be that the cost of avoiding these species is too great. An interesting role for dominance is discussed by Caldwell (1981) in the context of mixed species heron flocks. Other birds benefit from foraging with the dominant Snowy Egrets, whose dominance ensured that they had access to the best feeding sites in an ephemeral environment, hence other birds used Snowy Egrets' positioning as a signal of a good quality site on a larger scale. The Snowy Egret conversely was able to exploit the foraging of close-by subordinate species by supplanting them when they find good food patches on a much more local scale.

In some circumstances, one species may benefit from prey flushed by another. This has been observed in mixed species primate groups, and birds and/or marmosets associating with army ants (Peres 1993). This need not necessarily involve a cost to individuals of the 'beating' species, as the flushed individuals may be effectively lost to them anyway, or of no interest as a food source. Indeed, the beaters may derive anti-predatory benefits from the satellites. This has been especially observed in mixed species avian flocks, where the evolution of complex interspecific relationships involving specialization as sentinels may have developed, with the use of false alarm calls to obtain prey items (Munn 1986). This study described the behaviour of two species of fly-catching birds that lead mixed species foraging groups, act as sentinels, and feed mainly on insects falling through the air having been disturbed by other group members. Sometimes when diving after such a food item at the same time as another species, the sentinel species will emit a call similar to that used normally to warn off hawks. The effectiveness of this as a means of distracting the other bird was not reported.

6.6 Consequences of inter-individual differences for optimal group size

Intrinsically different individuals will often be influenced differentially by changes in group size. For example, increasing the size of a foraging group is likely to have a detrimental effect on subordinate individuals, but may have a positive effect on a dominant one. Hence, intrinsically different individuals are likely to have different optimal group size preferences. One might suspect that this will be a strong pressure driving assortativeness in groups, but Ranta *et al.* (1993) demonstrated that the situation is likely to be more complicated, since individuals are likely to 'disagree' not only on optimal group size but on optimal group composition. For example, a dominant individual may want to be in a large group but may also want that group to be composed mainly of subordinates, whereas subordinates may have a preference for a small group of their own. Ranta suggested that if changing groups comes at a low cost, then there will be fluidity of group size and composition, with individuals constantly moving, and groups constantly being created, changed, and dissolved. When opportunities for movement are less or the costs are higher, then the situation becomes more complicated, and general rules are hard to specify. However in many cases, where different phenotypes disagree on optimal group size and composition, the resulting groups will often not be optimal for any of their members.

6.7 Summary and conclusions

From a wide range of taxa we have empirical evidence that individuals preferentially choose to group with some over others. We also have considerable conjecture and some evidence for benefits that can arise from such non-random assortment. However, what is currently lacking is an understanding of how assorted groups form and of the qualitative costs and benefits to individuals of being in groups of different compositions. Such exploration requires a solid theoretical foundation, the predictions of which could then be used to identify critical empirical tests. The pioneering work of Ranta described in Sections 6.2 and 6.6 may provide a useful platform for a generalized theory of group composition. Alternatively or additionally, the system level modelling described in Chapter 9 may also be a fruitful line of approach. Either way, if we are to advance our current patchy knowledge of the mechanisms that underlie the common observation of assortment in nature, such theory is urgently needed.

7

Evolutionary considerations

7.1 Introduction

This chapter investigates how grouping has evolved in animals. The preceding chapters, in particular Chapters 2 and 3, provided information on the effects of grouping on factors such as foraging efficiency and predation risk of individuals. Increasing group size has the potential for reducing predation risk and for increasing foraging efficiency (up to a certain threshold group size). What is the evidence then that these conditions select for grouping behaviour over evolutionary time? And in what way do they shape the social organization of animal groups? Before we can address these questions we have to deal with another issue. It is not enough to show that grouping would be advantageous under certain ecological conditions. The latter only defines the evolutionary scenario, the starting conditions that have to be in place for grouping to evolve, but does not address the actual selection mechanism. For grouping to evolve by natural selection we have to show that there is variation in grouping tendency between individuals in a population and that grouping is a heritable behavioural trait that can be passed on from one generation to the next.

In the first part of this chapter, we will provide some experimental evidence that grouping (as a behavioural trait) can indeed be selected for (Section 7.2). Having addressed the mechanism, we will discuss how it operates in nature. Section 7.3 will review the evidence for population differentiation within species on the basis of eco-logical factors. The third part of this chapter (Section 7.4) will compare closely related species to identify the main ecological factors and life history traits, their specific role, and potential interactions with regards to the evolution of group-living in diverse taxa (Table 7.1). Having looked at the factors that potentially selected for group-living in individuals, we then consider whether groups, once they are in existence, can in themselves become units of selection (Section 7.5).

7.2 Individual differences: artificial selection

Ruzzante and Doyle (1991, 1993) carried out a number of artificial selection experi-ments with medaka, a small cyprinodontid fish. Individuals were selected for either fast or slow growth. Selection for fast growth rates over just two generations led to a decrease in aggression and an increase in shoaling tendency (Fig. 7.1). Aggression was found to be detrimental for a fish's growth rate, since energy was wasted on

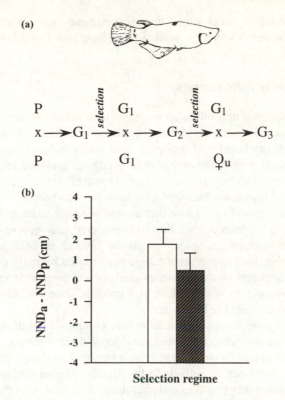

Fig. 7.1 (a) Selective breeding of Japanese medaka. Thirteen pairs of medaka were used to produce the first generation G_1 consisting of 95 broods. A total of 54 pairs from the G_1 which had been selected for fast or slow growth were used to produce the G_2 consisting of 182 broods. G_2 medaka were again selected for fast and slow growth and 31 males mated with unselected females to produce the G_3 giving 219 broods. (b) Mean ($\pm$ S.E.) shoaling response in medaka from G_3, which were selected for fast growth (open bars) or slow growth (hatched bars). The nearest neighbour distance between fish during predator presence (NND_p: fast growth = 6.1 cm, slow growth = 7.8 cm) was subtracted from that when a predator was absent (NND_a: fast growth = 7.9 cm, slow growth = 8.3 cm). (Adapted from Ruzzante and Doyle 1993.)

trying unsuccessfully to monopolize a food source in an environment where food was not limited. Ruzzante and Doyle's experiments support the idea that there is a genetic relationship between growth rate and social behaviour. The fact that a decrease in aggressive behaviour was correlated with an increase in shoaling tendency is intriguing and suggests that the two behaviours might potentially be genetically linked (however, this requires further investigation).

Selective breeding experiments are very useful in that they directly address the mechanisms that bring about change in grouping behaviour over time, and thus illustrate the potential for evolutionary change. While the information regarding the mechanism is valid, the conditions under which these experiments are carried out are often highly artificial, which is why they should ideally be coupled with field experiments.

7.3 Population differences

For grouping behaviour to result in an increase in individual fitness, it has to either increase birth rates, decrease the probability of death, or both. Chapter 2 discussed various anti-predatory benefits of grouping. This might make one expect that predation will eventually lead to the evolution of grouping, provided that it is a source of mortality of sufficient importance for prey. However, this assumption requires a leap of faith. How important does predation have to be to become a selective agent for grouping? And how do we know that an anti-predator strategy such as grouping will evolve as a consequence? Alternatively, prey may evolve morphological defences such as camouflage, spines, or poison instead, or behavioural defences other than grouping, such as increased refuge use or immobility (in connection with camouflage). Which anti-predator strategy evolves will largely depend on the ecological conditions, and on whether there is a predisposition towards one particular strategy based on existing life history traits.

At this point it is important to remember that grouping in itself does not always necessarily provide protection from predators, particularly if the prey organisms are not very mobile and thus cannot benefit from a predator confusion effect. Grouping in caterpillars, for instance, is unlikely to be effective against bird predators unless some other defence strategy such as distastefulness is developed in conjunction with living in groups. In general, many species of social insects have evolved morphological defences including poisonous stings and distastefulness that are advertised to predators by bright warning coloration. The link between a conspicuous warning colour/pattern and prey defences on the other hand is more likely to be learnt by predators (and thereby benefit prey) if prey occur at high densities and live in groups (see Section 2.4.5). Thus different defence strategies of prey often evolve in conjunction with each other (see also Chapter 8 on locust behaviour).

In the section below, we give examples of two ways of testing whether predation pressure will indeed lead to the evolution of grouping. A direct test would involve experimental manipulation of the predation regime of an animal population. Depending on the generation time of the prey, this can be a rather lengthy exercise. Alternatively, one can compare different populations of a species that exhibit natural variation in predation pressures. The latter means that no manipulation is necessary, but potentially confounding factors have to be dealt with, because different populations are likely to differ in more than just the intensity of predation.

7.3.1 Testing for population differences

As Huntingford *et al.* (1994) pointed out, variation regarding a particular behavioural trait can be partitioned into within-individual variation and between-individual variation. The former can be a result of changes experienced by the individual over time due to internal factors (e.g. state-dependent, parasite-meditated, or development-related changes) or external ones (e.g. environmental conditions, such as temperature changes). Variation between individuals can be due to differences between the sexes or other traits. Individual traits can be the result of phenotypic plasticity and learning via cultural transmission (Laland and Reader 1999), genetically-based changes in behaviour, or combinations of these factors (Ruzzante and Doyle 1991; Tulley and Huntingford 1987a,b).

Population differentiation in anti-predator morphology and behaviour has been described for many different species: small mammals (Towers and Coss 1990; Loughry 1988), garter snakes (Arnold and Bennett 1984; Herzog and Schwartz 1990), salamanders (Dowdey and Brodie 1989), freshwater fish (Magurran *et al.* 1992; Huntingford *et al.* 1994), and spiders (Riechert and Hedrick 1990). Studies of population differences in grouping behaviour, however, are rare, and come mostly from freshwater fish: minnow, *Phoxinus phoxinus* (Magurran 1986; Magurran and Pitcher 1987), and guppy, *Poecilia reticulata* (Seghers 1974; Magurran and Seghers 1991). In both guppies and European minnows, population differences in grouping behaviour persisted in laboratory-reared offspring, which demonstrates that these differences were not based on learning and probably had a genetic basis (Seghers 1974; Breden *et al.* 1987; Magurran 1990) (Fig. 7.2). However, work on threespine sticklebacks, *Gasterosteus aculeatus* (Tulley and Huntingford 1987a,b) showed that exposure to paternal care had a major influence on the expression of anti-predator behaviours of fish from high risk sites. Thus genetic predisposition and early experience in life can be closely linked.

7.3.1.1 Guppy shoaling behaviour

Magurran and Seghers's work on the guppy has provided major insights into how population differences in grouping tendency can arise and be maintained. It also gives an indication of the time scale over which these processes take place. Shoaling is one type of anti-predator behaviour of the guppy that is particularly effective against fish predators (Krause and Godin 1995). Seghers (1974) demonstrated a relationship between variation in the level of predation risk and shoaling behaviour in five populations of wild guppies. Tests of laboratory-reared fish demonstrated that guppies from high predation sites exhibited a stronger shoaling tendency than ones from low predation sites. This result was further supported by a more extensive study on the shoaling behaviour of nine populations of female guppies (Fig. 7.2) (Magurran and Seghers 1994). Magurran and Seghers (1991) argued that the variation in shoaling tendency between populations represents a behavioural adaptation to different environments. High predator density selects for strong shoaling behaviour, whereas low predator density favours the evolution of aggressive and territorial behaviour.

In 1957, Haskins moved 200 guppies from a high predation site to a low predation one. A similar transplant involving 200 fish was carried out by Endler in 1976.

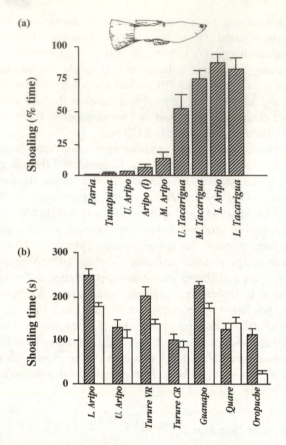

Fig. 7.2 (a) Variation in the percentage of time (+95% confidence limits) spent shoaling by wild female guppies from nine populations in Trinidad. (Adapted from Magurran *et al.* 1995.) (b) Significant variation between guppy populations persisted after breeding the fish under controlled conditions in the laboratory and testing their descendants (one-way ANOVA: females $F_{6,141} = 14.9$, $p < 0.001$; males $F_{6,141} = 22.7$, $p < 0.001$). Females (hatched bars) also showed a greater shoaling tendency than males (open bars), apart from the Quare population. Mean ($\pm$ S.E.) shoaling time with a group of six stimulus fish is given. (Adapted from Magurran *et al.* 1992.)

Analyses of the behaviours and morphology of the transplanted populations showed that morphologically both populations underwent significant changes (Reznick *et al.* 1990) but behaviourally only the fish from the earlier transplant had changed; decreasing their tendency to shoal (Fig. 7.3) (Magurran *et al.* 1992, 1995). The time period of 34 years between the first transplant and the subsequent test of fish shoaling behaviour is equivalent to approximately 100 guppy generations, whereas the second transplant allowed for only 30–40 generations. It is tempting to conclude that 30–40 generations simply was not long enough for shoaling behaviour to detectably

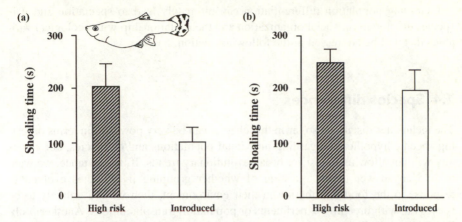

Fig. 7.3 (a) Mean (+95% confidence limits) shoaling tendency in female guppies that were moved from a high to a low predation (open bar) site by Haskins in 1957 and tested in 1991 by Magurran *et al.* (1992). The control fish (hatched bar) were taken from a high predation site further downstream and showed a stronger shoaling tendency than introduced fish. (b) Mean (+95% confidence limits) shoaling tendency in female fish that were moved from a high to a low predation site (open bar) by Endler in 1976 and tested in 1992 by Magurran *et al.* (1995). No significant difference in shoaling behaviour was detected between introduced fish and individuals from the original high risk site. (Adapted from Magurran *et al.* 1992 and 1995, respectively.)

change. However, Ruzzante and Doyle (1991, 1993) have shown that a few generations may be sufficient for behavioural change in artificial selection experiments, so the potential for rapid change probably exists. Although the same number (i.e. 200) of fish was transferred in each case, it is possible that there were great differences in the effective population size due to high mortality after the transfer or as a result of sperm storage and multiply inseminated broods (Magurran *et al.* 1995). Allozyme analysis, however, showed little difference in genetic variation between the introduced fish and the founder population, which suggests that the introduced fish flourished in their new environment. Furthermore, subtle differences in shoaling behaviour have been found between populations that are subject to similar levels of predation (i.e. have a similar predator community and density), suggesting that other factors (such as resource competition) also play a role.

Attempts have been made to link environmental and genetic variation in different guppy populations and to monitor manipulated populations to study the genetic consequences of transplants. Correlations between environmental and genetic variation are well supported. Allozyme analysis showed substructuring of populations that could be related to physical barriers and distance between sites. However, no direct relationship between predator regime and genetic diversity has been found, although this could be due to confounding factors such as drift and founder effects (Shaw *et al.* 1991, 1994).

Increasing population differentiation can ultimately lead to speciation and it is species differences in social organization and their relationship with ecological variables that will be addressed in the following section.

7.4 Species differences

The techniques discussed so far in this chapter can be very powerful in terms of testing specific hypotheses regarding individual populations and/or species, but generally will not allow us to test for broad evolutionary trends. If, for instance, we want to find an answer to the question of whether grouping in primates evolved in response to the food distribution in their environment, then we are unlikely to be successful with breeding experiments or population transplantations. Another problem is that not all populations are amenable to these approaches because of ethical or practical considerations. Transplants of locally adapted populations into new environments, for instance, pose a number of conservation problems. Furthermore, generation times can be rather long in some species and waiting to detect evolutionary change might occupy several generations of researchers.

An alternative approach is the comparative method, which usually involves a comparison between a group of different but closely related species concerning the relationship between their social organization and environmental variables (Harvey and Pagel 1991; Krebs and Davies 1993). The comparative method provides a powerful tool to help understand how species have adapted to their environment. The investigation starts by identifying a taxonomic group and by accumulating information on the life history and ecological variables of interest for each of the respective species. Some of the first comparative studies of grouping by Crook (1964) on weaver birds (see also Crook and Gartlan 1966 primates), and Jarman (1974) on African ungulates, are still unrivalled for the detail they provide on the natural history of each species. Since then, however, the comparative method as an analytical tool has undergone some drastic changes that introduced much needed statistical rigour (Clutton-Brock and Harvey 1977; Harvey and Clutton-Brock 1985; Harvey and Pagel 1991), and some of the older studies would probably benefit from re-analysis using current methodology (explained in detail by Harvey and Pagel 1991). A good example of this is a recent re-analysis of Jarman's (1974) data set by Brashares *et al.* (2000), who tested Jarman's hypotheses that:

(a) Dietary selectivity is negatively correlated with body mass and group size.
(b) Anti-predator strategies are group size dependent (i.e. group-living species flee when under predatory attack whereas solitary or pair-living species hide).
(c) Group size and body mass are positively correlated.

Brashares *et al.*'s (2000) approach took the phylogenetic relationship of the 75 species of African ungulates into account following the phylogeny by Gatesy *et al.* (1997) (Fig. 7.4a). Their analysis supported the hypothesis that species with a selective diet

were found in smaller groups and had lower body weights than unselective species
(Fig. 7.4b). Smaller ungulate species such as dikdik and steenbok are more likely to
be browsers on highly nutritious plants, rather than unselective grazers, because of
their higher mass-specific energy requirements. Grasses tend to have a more even
distribution than shrubs, which means that competition for food among group members should be more likely in browsers compared with grazers, which in turn reduces
group sizes. Brashares *et al.*'s analysis also supported the hypothesis that ungulates
in open habitats were more likely to live in groups and adopt flight as an anti-
predator behaviour compared with solitary species that live in bush or woodland, and
tend to hide when attacked by a predator (Fig. 7.4c). This difference in anti-predator
behaviours was significant even when body size was included as a covariate in the

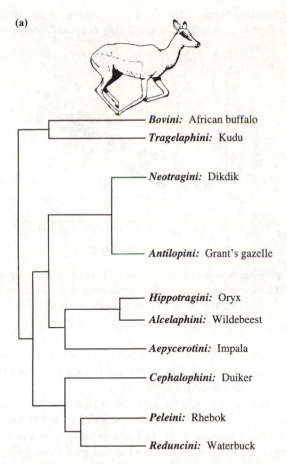

(a)

Bovini: African buffalo
Tragelaphini: Kudu
Neotragini: Dikdik
Antilopini: Grant's gazelle
Hippotragini: Oryx
Alcelaphini: Wildebeest
Aepycerotini: Impala
Cephalophini: Duiker
Peleini: Rhebok
Reduncini: Waterbuck

Fig. 7.4 (a) Phylogenetic tree of the 10 major taxonomic groups of African ungulates
(adapted from Brashares *et al.* 2000). The English names of some of the commonly known
species belonging to these groups are given.

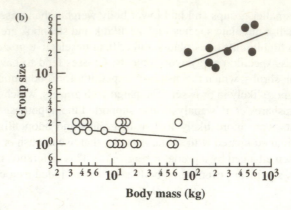

Fig. 7.4 (b) Relationship between diet selectivity (selective browsers: open circles, and unselective grazers: filled circles), body mass, and group size.

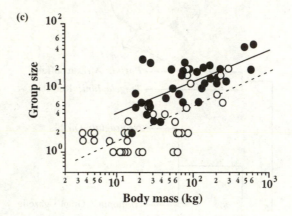

Fig. 7.4 (c) Relationship between group size and body mass for antelope species that live in bush and forest habitats and hide (open circles) in response to predators and ones that live in open habitats and flee (closed circles).

analysis. No clear support was found for a positive relationship between body mass and group size, although this could be partly attributable to lack of statistical power when the analysis was corrected for phylogeny.

In conclusion, the results indicate that the differences in group size patterns between ungulate species in open areas and in woodland are not simply a by-product of body size differences but reflect, at least in part, adaptations to different habitats. The fact that ungulates in open areas have little to gain from grouping in terms of increased foraging efficiency (e.g. increased probability of food patch detection) suggests that larger group sizes have primarily evolved as an anti-predator strategy. However, this idea needs critical testing.

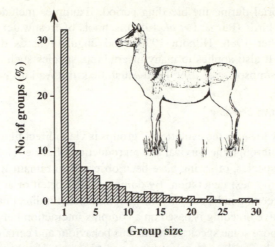

Fig. 7.5 Group size distribution of impala: $n = 1303$ groups, median group size $= 3.1$, mean group size $= 6.8$. (Adapted from Wirtz and Lörscher 1983.)

While the requirements for carrying out comparative analyses are clear (see Harvey and Pagel 1991), the application is often less straightforward for a number of reasons, including incomplete data sets on life histories and problems with the reconstruction of phylogenetic trees. A weakness of Jarman's data set, for instance, is that diet and habitat preferences of different species were recorded as categorical variables and not as percentage preferences, which makes the analysis regarding these factors less powerful. Furthermore, mean group sizes were used, when the median is probably a better indicator of group size distributions because of their skewed nature (Fig. 7.5).

7.4.1 Pathways towards the evolution of groups

From a mechanistic perspective, a preference for close spatial association with conspecifics has to evolve for groups to come into existence. Whether or not a mutation that possesses this grouping tendency (in a population of solitary individuals) will be successful and spread, will depend on the trade-off between the costs and benefits of grouping. The latter can only evolve if there are net-benefits to the grouping strategy compared with the solitary one, resulting in higher breeding success, and thus direct fitness, of grouping individuals (Vehrencamp 1983b; Turner and Pitcher 1986). This is the pathway leading towards the evolution of groups of non-related individuals whose group composition is rather fluid, being characterized by so-called fission–fusion processes which allow for free exchange of individuals between groups: although a certain degree of assortment by phenotypic traits, e.g. size- and species-assortedness (Krause *et al.* 2000a,b) is common. Many species of vertebrates fall into this category, only forming groups outside the mating season, with individuals becoming territorial during the breeding period. Examples include mixed species winter flocks of birds (Barnard *et al.* 1982), schools of freshwater and marine fish (Allan and Pitcher 1986; Hilborn 1991), and ungulate herds during migration (Sinclair 1977). It also applies to many invertebrate species such as mosquito and locust swarms (Simpson *et al.* 1999) and Antarctic krill (Watkins *et al.* 1992).

7.4.1.1 Co-operative breeding

Another pathway towards the evolution of groups is via indirect fitness (i.e. a genetic benefit is gained through the survival and reproduction of close relatives). In a number of different species, offspring have been observed to remain with their parents helping to raise the next generation. By doing so they forfeit or at least delay their opportunity of breeding independently but gain in terms of indirect fitness. The evolution of co-operative breeding is based on a complex interaction between life history traits that predispose some species towards this behaviour and certain ecological factors that have to be in place to select for it and trigger it. Hatchwell and Komdeur (2000) listed four major ecological constraints on independent breeding:

1. An absence or shortage of suitable breeding habitat.
2. Reduced survival probability following dispersal.
3. Reduced probability of finding a mate.
4. Reduced chance of successful breeding once a territory has been established.

The limitation of suitable breeding habitat was the first explanation put forward for the evolution of co-operative breeding (Selander 1964) and has since met with good empirical support. It has been shown that habitat saturation can be directly responsible for the occurrence of co-operative breeding in the Seychelle's warbler, a small passerine bird (Komdeur 1992). Furthermore, offspring that remain with their parents have a high probability of taking over their parents' territory at some point, which would bring a direct fitness benefit (Woolfenden and Fitzpatrick 1978).

In marmots, small to medium size mammals, social systems range from solitary species (such as the woodchuck) to family groups comprising the dominant pair, subordinate adults, yearlings, and young offspring (e.g. alpine marmot). Offspring remain as helpers with their parents in species where the growth period during the summer is too short to attain full body size (Armitage 1999). Delaying dispersal brings the benefits of reduced predation risk and lower winter mortality. Lower predation risk has also been reported as a factor promoting coloniality in the spider genus *Anelosimus*, where solitary species, periodically social ones, and permanently social ones can be found. In the periodically social species, the offspring stay in the mother's nest for almost a year and disperse shortly before the next breeding season (Fig. 7.6). In permanently social species, several generations live together in the same nest and matings take place between members of the same colony (nest). How can we imagine that the transition from solitary spiders to permanently social ones took place?

Aviles and Gelsey (1998) suggested that one of the crucial factors for the evolution of sociality in spiders is increased offspring survival in communal nests.

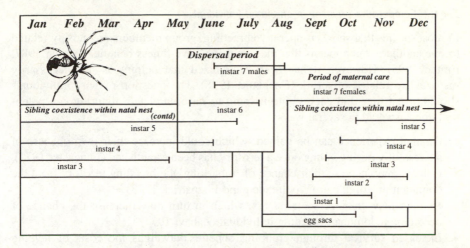

Fig. 7.6 Seasonal dispersal in the periodic social spider, *Anelosimus jucundus*. (Adapted from Aviles and Gelsey 1998.)

However, a continuous supply of prey items is required for the upkeep of communal nests, which is probably why colonial spiders can only be found in the tropics where insects are available year round. It has been demonstrated that the timing of dispersal can be delayed in periodically social spiders if additional food is provided. This suggests that competition plays a role (Krafft *et al.* 1986; Gundermann *et al.* 1993). Another important function of dispersal, apart from avoidance of competition, is avoidance of inbreeding. However, if prey densities are high and competition within the colony low, then high survival probabilities may favour life in the colony even at the cost of inbreeding depression. It seems likely that recessive deleterious alleles will be selected against in a population over evolutionary time, which might make the existence of completely inbred social spiders possible (Riechert and Roeloffs 1993). Although the above line of argument for the evolution of sociality in spiders seems plausible, more experimental work is necessary to test alternative explanations. Given the short generation time of spiders and the easy access to colonies for experimental work in both the laboratory and the field, this system is probably one of the most promising ones for further investigation.

Pruett-Jones and Lewis (1990) carried out an elegant experiment with fairy-wrens, small passerine birds, which demonstrates the importance of mate availability. If the breeding male was removed from its territory but the breeding female was left undisturbed, then a non-breeding male, that was a helper on a nearby territory, would establish itself within a few hours as the new territory holder. If, however, both male and female were removed, dispersal of helpers was not observed for several days and indeed only took place after the female had been released. This indicates that mate shortage was more important than territory availability in this case.

7.4.1.2 Co-operative breeding in unrelated individuals

In most species that show co-operative breeding, group members are closely related to one another, which means that they gain inclusive fitness benefits (Brown 1987; Emlen 1995). However, it has been recognized that helping is not necessarily restricted to related offspring (Cockburn 1998). The question of why individuals should invest time and energy into helping to raise unrelated offspring can potentially be answered in several ways:

1. Parental experience can be gained, which could increase their own chances of successfully raising young once a territory has been established (Lancaster 1971).
2. Helpers may increase their chance of 'inheriting' the breeding territory from the dominant breeding pair (Woolfenden and Fitzpatrick 1978).
3. Helping may enhance social status, which in turn may increase the chance of becoming a dominant breeding individual (Zahavi 1977).
4. Increased survival through grouping benefits outweighs the costs of helping (Wiley and Rabenold 1984; Kokko et al. 2000).

Furthermore, if a subordinate helper eventually becomes dominant, then it would 'inherit' the unrelated helpers, which would then increase its own reproductive success.

7.4.1.3 Life history traits

Co-operative breeding is not simply a result of the right ecological conditions but also requires the existence of certain life history traits that have been recognized as important predispositions for co-operative breeding. The occurrence of co-operative breeding is thus dependent on the evolutionary past as well as the current ecological conditions. Co-operative breeding is generally associated with populations where high population density and intense competition in a stable environment have selected for deferred maturity and the production of small numbers of offspring that receive high degrees of parental care. In a comparative study of 182 co-operative and non-co-operative species of birds, Arnold and Owen (1998) reported that co-operative breeding was associated with high adult survival, low dispersal, and small clutch sizes. A general problem with comparative studies is that they do not identify which is cause and which is effect. Therefore factors such as high adult survival, low dispersal, and small clutch size could also be consequences (rather than causes) of co-operative breeding. However, for at least one of these factors the authors could show some tentative evidence for a causal relationship. Annual mortality was found to be lower in species that do not breed co-operatively, but belong to families that have a high proportion of co-operative breeding species, compared with families with a low proportion of co-operative breeders. Therefore, they concluded overall that the main factor that predisposed certain avian families to co-operative breeding is low annual mortality.

The importance of life history traits for the evolution of sociality is further exemplified by the eusocial insects that form some of the most complex social systems in the animal kingdom. Eusociality, which is mainly found in the hymenoptera and termites but has also been described for a number of aphids (Aoki 1982), beetles (Kent and Simpson 1992), thrips (Crespi 1992), and spiders (Vollrath 1986), is generally defined by the presence of sterile castes, co-operative brood care, and the coexistence of three or more different generations in the same colony. Two different pathways towards the evolution of eusociality have been suggested (Fig. 7.7). The first one, the so-called subsocial route, is via co-operative breeding and similar to the above described pattern. The offspring remain at the nest and provide protection for the next generation, which are their siblings. The high degree of relatedness is an important genetic predisposition for the evolution of helping behaviour in this case. The helpers at the nest do not reproduce themselves and in the more advanced eusocial species form sterile castes. The second pathway is the parasocial one, which starts with several females sharing a nest and the care of offspring. There is some dispute over the degree of relatedness in these females, which in most cases are sisters (Crozier and Pamilo 1996). At the next stage of the process one female begins to dominate the others, which then stop reproducing and become workers.

An important factor predisposing the hymenoptera for the evolution of eusociality is the fact that males are haploid and females diploid (Hamilton 1964). This

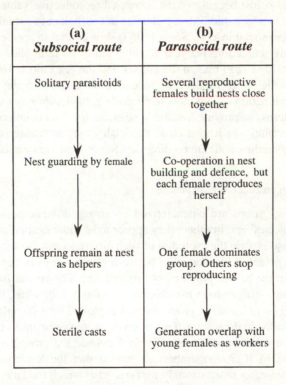

(a) *Subsocial route*	(b) *Parasocial route*
Solitary parasitoids ↓	Several reproductive females build nests close together ↓
Nest guarding by female ↓	Co-operation in nest building and defence, but each female reproduces herself ↓
Offspring remain at nest as helpers ↓	One female dominates group. Others stop reproducing ↓
Sterile casts	Generation overlap with young females as workers

Fig. 7.7 Different pathways towards the evolution of eusociality: (a) subsocial route, (b) parasocial route. (Adapted from Krebs and Davies 1993.)

results in females being more closely related to sisters than to daughters, which favours co-operative brood care over raising their own offspring. However, as Trivers and Hare (1976) pointed out, the low relatedness of females to their brothers poses a problem. If the queen produces equal numbers of males and females, then the fitness of an average non-reproductive female helping at the nest is the same as that of a reproductively active female, and therefore no incentive exists for the evolution of helping. To maximize their fitness female workers should manipulate the sex ratio in favour of females (3:1 reproductive females to males) in their colony. A skewed sex ratio, however, is not in the interest of the queen which is equally related to sons and daughters (see Crozier and Pamilo 1996 for a detailed discussion of this topic). The outcome of this conflict depends on who is in control in the colony. Trivers and Hare (1976) found that in 21 species of ants, where the queen mated with only one male, the sex ratio in the colonies was closer to 3:1 than to 1:1. However, there are also examples of colonies where the queen appears to be in control (see Crozier and Pamilo 1996, Chapter 5).

It has been pointed out that a sex ratio bias of 3:1 in favour of females would only provide an incentive for workers to be helpers, if the sex ratio in the population is 1:1 (Seger 1991). Otherwise the genetic benefit to helpers due to close relatedness with their sisters is lost because of the increased reproductive value of males. This argument explains why haplodiploidy alone is not sufficient to explain the evolution of sterile castes in social insects. Seger (1983) showed that in species with two partially overlapping generations per year and alternating sex ratio biases, the difference between the local sex ratio bias in the colony and the sex ratio in the overall population can be sufficient to produce an incentive for helpers to stay in the colony. Therefore a combination of factors is required (e.g. haplodiploidy as well as certain life history patterns) to provide a suitable scenario for the evolution of eusociality. In contrast, in termites which are diploid, eusociality may have arisen from an alternation between inbreeding and outbreeding (see Seger 1991 for details).

7.4.1.4 Social organization: hierarchies

In some species, groups are characterized by strong dominance hierarchies (e.g. some primate species) and in others they appear to be more egalitarian (e.g. shoaling fish). How can we explain the evolution of such differences in the social organization of groups? The potential for inequality between group members depends largely on the magnitude of the benefit potential of group-living. The greater the fitness difference between an average group member and a solitary individual, the less likely a subordinate will be to leave the group, even if exploited by a dominant group mate. The dominant individual can bias the resources in its favour up to the point where subordinate group members fare equally well outside the group (Alexander 1974; Vehrencamp 1983b). If group members are related, then the dominant individual can exploit the subordinates more strongly to the level at which the latter's direct fitness is below that of solitary conspecifics. However, the subordinates' inclusive fitness cannot decrease below that of solitary conspecifics, otherwise the group will disperse

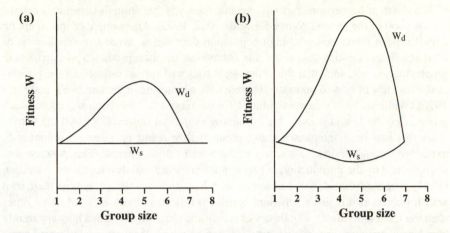

Fig. 7.8 Differences in direct fitness between dominant (W_d) and subordinate (W_s) group-members for groups of (a) non-relatives and (b) relatives. (Adapted from Vehrencamp 1983b.)

(Fig. 7.8). Recent debate has centred on the different strategies that dominants can adopt under these conditions. Crespi and Ragsdale (2000) suggested that dominants could manipulate their offspring to be small by means of harassment and withholding food. This would lower the chances of offspring survival outside the group and thus increase the likelihood of them staying to help. However, the authors conceded that parental manipulation would only be selected for after helping had already evolved. Another way in which a dominant can ensure the presence of helpers is to give them a share in the reproduction (Alexander 1974; Reeve and Ratnieks 1993). The prediction of reproductive skew (distribution of reproduction between individuals) within animal groups has been the focus of a number of empirical and theoretical studies in recent years (Clutton-Brock 1998; Kokko and Johnstone 1999).

However, there are also other limitations regarding the evolution of dominance relationships within groups, such as predation pressure. As outlined in Chapter 6, differential predation favours the evolution of phenotypic homogeneity in mobile groups when predators are affected by the confusion effect. Larger dominant individuals are prevented from invading groups of smaller conspecifics because of the oddity effect which increases their predation risk and thereby outweighs the gains resulting from physical superiority that allows the dominant individual to monopolize resources (Ranta *et al.* 1994; Peuhkuri 1997).

7.5 Groups as units of selection

This topic has been a very controversial subject in evolutionary ecology for decades, and very strong views abound (e.g. Maynard Smith 1998).

The classical suggestion of group selection theory is that animals sometimes behave for the good of the group (Wynne-Edwards 1962, 1993). An example of this might be curtailment of reproduction at high population densities to avoid overexploitation of food resources. Groups that show this behaviour are then predicted to outcompete groups that do not, such that this group level trait will spread throughout the population. This view of how to understand nature took a battering in the late 1960s and early 1970s (Williams 1966; Maynard Smith 1976). First of all, it has been argued that such group selection is unnecessary for explaining empirical observations. All the behaviours that had been interpreted using group theory could be more parsimoniously explained through individuals acting in their own selfish interest. This position was strengthened by the introduction of two important theoretical developments. Hamilton (1964) proposed the idea of kin-selection, where individuals act against their own selfish interest if this gives sufficient benefit to related individuals (that share genes with the focal individual). The theory of reciprocal altruism explained how apparently altruistic behaviour can evolve even in non-relatives through evolutionary game theory (Trivers 1971; Axelrod and Hamilton 1981). With these theories being widely accepted, there seemed no need for group selection. Further, there were theoretical arguments for why selection at a group level would be much less powerful than selection at an individual level: groups are generally longer-lived than individuals and variance between groups is likely to be less than between individuals, so evolutionary processes would run at a slower rate (Reeve and Keller 1999). However, if group selection is ineffective and unnecessary, why are we still talking about it?

In the first place, to some authors it seems that kin-selection and reciprocal altruism can be seen in a group selectionist context (Price 1970; Laland *et al.* 2000; Okasha 2001). Here we have to be careful about what we mean by group selection. The modern proponents of group selection prefer the term multi-level selection for their models (Sober and Wilson 1998), but the key point is also true for earlier group selection models: that these models do not represent selection in the same way as the verbal description given in the first paragraph of this section may suggest. They do not present group selection as a perfect analogue of individual selection, that simply operates at a higher level of organization. The reason for this is as follows: individual selection identifies the fittest individuals as those that contribute the largest number of offspring to the next generation and thus maximize their gene contribution to future generations. However, group selection does not identify the fittest groups as those that contribute the largest number of groups to the next generation. Rather both traditional group selection and modern multi-level selection identify the fittest groups as those that contribute the largest number of individuals to the next generation (Okasha 2001). Hence, like individual selection, the success of a group is defined in terms of the number of offspring produced or more specifically the gene contribution to future generations.

A discussion of the empirical evidence for (or against) group selectionist processes in nature is complicated by the ongoing debate over different models of group selection and their predictions, and uncertainty over what constitutes empirical evidence. In Wilson's (1980) model, a population fragments into groups of random composition that are subject to selection, resulting in some groups producing more offspring than others. During the selection period, movement of individuals between groups is restricted. This is an important requirement because group selectionist models are only relevant for the evolution of behavioural strategies whose pay-off is frequency dependent (i.e. dependent on the proportion of other group members playing the same strategy). Landeau and Terborgh (1986) provide a nice example of a frequency-dependent system in a predator–prey context. They studied the shoaling behaviour of blue and brown minnows, small freshwater fish, that were attacked by bass (a larger fish). If a single blue minnow switched from a shoal of blue minnows to one consisting of silver ones, this did not only dramatically increase its own predation risk due to the oddity effect. The presence of a blue minnow in a group of silver ones also decreased the overall predator confusion effect resulting in a higher predation risk for all group members.

Thus a behavioural trait needs to influence the fitness of other group members and not only that of the individual that carries it (to be of relevance to group selection). Another important point is that after a period of selection acting on individual groups, there has to be a general mixing of all individuals in the population followed by the re-formation of groups. This is a necessary requirement for any traits to spread in the population as a whole.

The discussion about potential empirical evidence for group selection has centred on a few taxonomic groups such as primates, viruses, and particularly social insects, but the evidence supporting group selectionist theory remains scarce and in part anecdotal (Wade 1976; Bradley 1999; Wilson 1997; Dugatkin 1997).

In conclusion, it seems that we are edging towards a consensus regarding the processes underlying the evolution of behaviour. Multi-level selection models seem not to be simply wrong, they are just another way of looking at aspects of natural selection that are less controversial when described in other ways. They are not describing a fundamentally different process of evolution, just a different way of representing the theory we already generally accept. The question we have to address is whether this alternative description makes understanding the natural world easier.

7.6 Summary and conclusions

Laboratory work has shown that grouping behaviour is a potentially heritable behavioural trait that can be selected for in artificial breeding experiments producing changes in grouping tendency within just a few generations in some species. This illustrates that the potential for rapid change in grouping behaviour exists in animal populations with short generation times. Manipulation experiments in the wild provide some evidence for a heritable reduction in grouping behaviour in the absence of predators. However, the time-span required for this change was considerably longer than expected from laboratory experiments. Nevertheless these field experiments convincingly demonstrated a link between ecological conditions and behavioural anti-predator adaptations. The genetic basis of such behavioural differences, however, is still little understood and remains a challenge for future studies.

Species comparisons allow far-reaching conclusions regarding the relationship between ecological variables and social organization and have been successfully employed to unravel evolutionary trends where direct testing has proved difficult or impossible. Despite a wealth of comparative analyses on social behaviours there are relatively few recent studies that have focussed on grouping behaviour and that have made use of the methodological advances of the last decade (exceptions are Arnold and Owen 1998, 1999; Brashares *et al.* 2000).

A particularly active area of research has been the topic of co-operative breeding in recent years. Why is co-operative breeding so interesting and relevant for the study of the evolution of grouping? In a given population it is only found in some individuals but not in others and its occurrence is dependent on the prevalent environmental conditions. Thus it is a temporary phenomenon which may disappear as soon as the conditions change. A comparison of different species also shows that even if the environmental conditions are favourable for the occurrence of co-operative breeding, only some species will actually show the behaviour. This has led to the identification of certain life history traits (such a low annual mortality) which predispose some species towards co-operative breeding. Finally, the fact that co-operative breeding is usually associated with a high degree of relatedness of group members means that it can increase the inclusive fitness even of non-breeders which allows for the evolution of unique social structures.

In conclusion, a number of theoretical models and empirical studies have greatly increased our understanding of the evolutionary pathways towards grouping behaviour. In particular, recent studies on group augmentation involving the recruitment of unrelated helpers look promising in that they begin to provide a conceptual framework which incorporates both the benefits of increased survival and foraging efficiency traditionally associated with grouping, as well as the advantages of increased breeding success (Table 7.1).

Table 7.1 Factors involved in the evolution of grouping behaviour

Factor	Approach	Study organism	Authors
Resource distribution	Selective breeding	Japanese medaka	Ruzzante and Doyle 1993
Predation risk	Transfer experiment	Trinidadian guppy	Magurran *et al.* 1995
Reproductive success	Comparative method	Co-operatively breeding birds	Arnold and Owen 1998, 1999
Resource distribution/ predation risk	Comparative method	African ungulates	Brashares *et al.* 2000

The first two studies were done on single species and provide direct evidence for heritable changes in grouping behaviour over several generations. The latter two studies inferred the role of ecological constraints and life history traits in the evolution of grouping from species comparisons. In birds, low annual mortality was identified as a potential predisposition for co-operative breeding whereas in African ungulates, body size and associated differences in metabolic rates are believed to be an important predisposing factor for the evolution of grouping.

8

Environmental effects on grouping behaviour

8.1 Introduction

We discussed in the preceding chapter how grouping behaviour can evolve in a population, and how the tendency for grouping can change over evolutionary time as an adaptation to environmental selection pressures. However, there are also many ways in which the behaviour of an individual animal can be affected by the environment during its lifetime. Phenotypic plasticity as a short-term response of the organism to its environment is particularly common during the early stages of development when tissues and organs and their nervous control are in the process of differentiation. Such plasticity should be specially selected for in changeable environments where conditions fluctuate within a generation. Using a number of case studies, we will first discuss the development of the sensory systems involved in grouping (Section 8.2) and the importance of ontogenetic shifts in sociality (Section 8.3), followed by a description of the influence of hormones on the expression of different phenotypes in hymenoptera and locusts (Sections 8.4 and 8.5). Another strategy that allows for adaptation to the environment during an individual's lifetime is a change in behaviour based on experience (Section 8.6). Learning is a particularly effective way in which organisms can deal with environmental changes on a short-term basis. Furthermore the content of what has been learned may not only benefit the respective individuals but can be passed on to conspecifics of the same or different generations. Finally, not all behavioural changes observed in animals are necessarily for their own benefit. Some parasite species have evolved elaborate ways in which to manipulate the grouping behaviour of their hosts for their own advantage (Section 8.7).

8.2 Ontogenetic constraints on grouping: fish shoaling behaviour

The mechanisms of grouping involve different sensory modalities in different species. Many terrestrial mammals use olfactory and visual cues for grouping, cetaceans use acoustic stimuli, birds rely mainly on visual ones, fish predominantly use a mixture of visual and mechanical stimuli (the latter are perceived via the lateral line), and many social-living insects dependent largely on olfactory information. In many vertebrate species, the sensory systems required for grouping behaviour are already fully developed at birth. An exception are many pelagic species of shoaling fish, which produce large numbers of planktonic larvae possessing only very limited swimming capacity, resulting in an inability to resist the currents in their environment. The onset of shoaling behaviour generally takes place during metamorphosis when the individual starts developing rods on the retina, red–white musculature, a gas-filled swim bladder, a more complex lateral line, and scales on the body (Gallego and Heath 1994; Fuiman and Magurran 1994). In herring, this change occurs when the individual is already several weeks old and about 35–40 mm long (Fig. 8.1). At this point in the development, the transparent larvae begin to show pigmentation of the eyes and growth of scales, making them increasingly conspicuous to predators. Hence, the development of shoaling behaviour is probably driven by an increase in predation pressure on individuals—a hypothesis that requires further investigation. The simultaneous development of the musculature provides the required capacity for positional control relative to conspecifics. This is an important prerequisite for shoaling in connection with increased differentiation of the lateral line and the eyes.

Recent work by Masuda and Tsukamoto (1998) on striped jack, a pelagic marine fish species, showed that another important factor (which has so far been neglected) could be the development of the central nervous system. This is essential for processing and co-ordinating visual and mechanical input and to produce mutual attraction between individuals and parallel orientation of fish. This view is supported by the fact that most sensory systems and locomotory organs required for shoaling behaviour were largely developed before mutual attraction in striped jack (leading to the formation of shoals) occurred: when body lengths of 12–16 mm were reached (Fig. 8.2).

The other extreme in the fish world in this context are live-bearers, such as the guppy (*P. reticulata*), where a fully developed juvenile is born and capable of shoaling almost immediately after birth (Magurran and Seghers 1990).

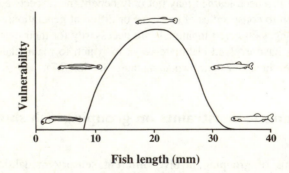

Fig. 8.1 Developmental stages of the herring and their vulnerability to predation. The development of scales during metamorphosis makes the fish more conspicuous to predators resulting in an increase in vulnerability. With a further increase in size, herring attain higher swimming speeds and begin to form shoals as an anti-predator defence that reduces vulnerability again. (Adapted from Fuiman and Magurran 1994.)

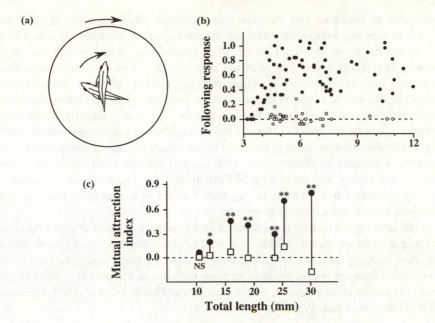

Fig. 8.2 (a) Measurements of the optokinetic response are used to test eye development. The circular tank was surrounded by a screen with vertical black and white stripes that was rotated and the following response of the fish measured. In control trials the screen was stationary. (b) Ontogeny of the circular optokinetic response in striped jack, *Pseudocaranx dentex*. Values (filled circles) above the dotted line indicate following behaviour significantly different from random movements. Controls (open circles) were run with a stationary screen. (c) Onset of shoaling behaviour as indicated by mutual attraction between fish. Three tanks were placed side-by-side. The test fish that were kept in the central tank, were presented with a stimulus shoal of conspecifics in one of the outer two tanks. The authors measured (filled circles) how much time the test fish spent on that side of their test tank that was closer to the stimulus shoal. As a control no stimulus shoal was presented (open squares). Significance is indicated by asterisks: * $p < 0.05$, ** $p < 0.01$. (Adapted from Masuda and Tsukamoto 1998.)

8.3 Ontogenetic shift in sociality: the spiny lobster

The temporal dynamics of grouping can be of great importance. For some organisms whether or not to group is an hourly or daily decision. For others the disbanding of groups is an annually recurring event, such as in many songbirds which form flocks during the cold months of the year but establish territories in spring and summer. In some species, like the spiny lobster, e.g. *Panulirus argus* and *Jacus edwardsii* (Ratchford and Eggleston 1998; Butler 1999), an ontogenetic shift in sociality can be observed. Juveniles are largely solitary and the tendency to aggregate becomes stronger as they grow: adults live mostly in groups. In contrast many other animal species have a greater tendency to group as juveniles when they are particularly vulnerable to predation and then disperse as adults, when competition costs are higher. In the spiny lobster, predation risk is also higher for juveniles compared with adults, but the juveniles do not benefit from grouping as an anti-predator strategy, unlike older conspecifics, and therefore rely on other forms of defence against predators, such as crypticity and refuge use (Eggleston *et al.* 1990; Butler 1999). Adult lobsters also use refuges as anti-predator protection but this strategy becomes particularly successful when practised in groups. The mechanism by which adult lobsters aggregate in crevices is largely based on olfactory cues. Experimental work has shown that lobsters are only attracted to the olfactory cues of conspecifics after reaching a certain size (Butler *et al.* 1999). Adult lobsters need sufficiently large crevices for hiding, and these may be difficult to find. The odour of conspecifics already positioned in such shelters may therefore act as a 'sign-post' towards such locations as well as facilitating grouping.

It has been argued that the usual trend of group-living as a juvenile with a decrease in social tendencies towards adulthood should be reversed in species where juveniles are slow moving and largely defenceless against predators and rely mainly on crypticity. Grouping would result in making individuals more easily detectable to predators and increase predation risk in such cases (Tinbergen *et al.* 1967; Treisman 1975; Dukas and Clark 1995; Butler *et al.* 1997).

8.4 The role of rearing conditions: caste determination and division of labour in the honey bee

In social insects, caste determination of females (i.e. non-reproductive as a worker or reproductive as a queen) is largely under the control of the workers, and thus due to rearing conditions. In honey bees, fertilized diploid eggs develop into females and unfertilized, haploid eggs into males. Whether or not a diploid egg develops into a worker or a queen, however, depends entirely on the rearing conditions. If the egg is laid into a queen cell, the quantity and quality of food provided by workers to the larva will be different from that received by a larva in a worker cell. Different feeding regimes affect the phenotype via the endocrine system, with higher levels of juvenile hormone leading to the development of a queen (Fig. 8.3). Exchanges of eggs between worker and queen cells have shown that development is exclusively under environmental control.

Juvenile hormone is also thought to play an important role in determining temporal caste structure. The life of a worker can roughly be divided into four phases (between which there can be considerable overlap). The first three phases involve tasks performed inside the hive such as the cleaning and capping of cells, the rearing of larvae, comb building, and food handling from returning foragers. The final phase involves guarding the nest entrance as soldiers and carrying out foraging trips outside the nest. Given the temporal succession between these phases, younger bees tend to be more active inside the hive and older ones outside it. However, the

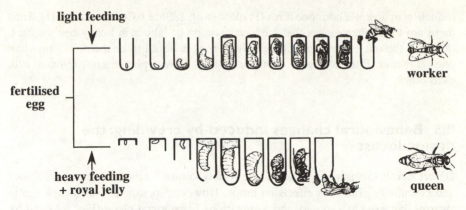

Fig. 8.3 Phenotypic plasticity of the honey bee. A fertilized egg can develop either into a worker (upper panel) or into a queen (lower panel) depending on the quantity and quality of the food provided by the workers. (Adapted from Gullan and Cranston 2000.)

correlation of task performance with age is weak and at least partly influenced by environmental factors. If the density of older foraging bees is low and food in the colony runs short, accelerated endocrine and behavioural development ensure that some of the younger workers start foraging precociously to replenish food stores (Robinson 1992; Schulz *et al.* 1998). If, in contrast, a shortage of young workers arises, so that not enough individuals are available for rearing larvae and cleaning the nest, then older workers regenerate their wax glands and revert to performing duties inside the hive. By de-coupling the level of food stores of a colony from the nutritional status of an individual worker, Schulz *et al.* (1998) were able to show that division of labour was not mediated by a worker–nest interaction. They suggested that chemical signals that are exchanged between workers during mandibular contact involved in food exchange (a widespread activity among workers) are responsible for stimulating the production of the hormones that control division of labour (Huang *et al.* 1998). A number of studies have shown that the behavioural development is influenced by neurotransmitters, e.g. dopamine, serotonin, and octopamine (Schulz and Robinson 2001), and juvenile hormone titres that increase throughout an individual worker's life. Experimental removal of the glands that produce juvenile hormone resulted in a delayed onset of the foraging phase (Sullivan *et al.* 2000). Older bees are thought to produce a (yet unidentified) substance that inhibits the development of younger workers and which is passed on during mandibular contact. Thus the presence of older bees inhibits biosynthesis of juvenile hormone in younger ones, regulating the proportion of individuals working inside and outside the hive (Huang *et al.* 1998). The mandibular pheromone of honey bee queens and brood pheromone (located on the cuticles of honey bee larvae) have both been shown to produce a similar delay in development of workers via a reduction in juvenile hormone titres (Pankiw *et al*. 1998; Le Conte *et al*. 2001). Thus there are three pheromones that mediate division of labour in honey bee workers, of which the one exchanged between workers is believed to be the most important one. However, little is known about how the different pheromones interact with each other.

8.5 Behavioural changes induced by crowding: the desert locust

Desert locusts (*Schistocera gregaria*) are widely known for their swarming behaviour and potentially disastrous effects on crops. However, locusts do not always live in swarms, but exist in a solitary and a gregarious behavioural phase (Fig. 8.4a and b). Phase changes can occur during an individual's lifetime but have also been demonstrated to be inter-generational.

Under conditions of high density, visual and olfactory stimuli and, to an even greater extent, tactile ones induce a tendency towards gregarious behaviour in locusts within a matter of just one hour, becoming fully expressed after 4–8 hours (Fig. 8.4) (Simpson *et al*. 1999). It is assumed that such effective short-term modification of behaviour is achieved by biochemical reinforcement of synaptic connections. Recently, the main site of mechanosensory input that elicits the transition to gregarious behaviour has been identified as the back legs, i.e. hind femurs (Simpson *et al*. 2001). The reason is probably that the hind legs are ideally positioned for the detection of conspecifics around them. In contrast, other body parts such as the mouth parts, tarsi, and lateral thorax are often stimulated during normal foraging and locomotion and therefore would be less likely to provide unambiguous information about locust density.

Longer-term crowding of locusts results not only in increased behavioural tendencies for aggregation, but also in a change of morphology, both of which are under hormonal control. The solitary form is green whereas the gregarious one is yellow and black. The colour change, which can be described as a continuous polymorphism, since it happens gradually, is in part brought about by neurosecretory cells in the brain which produce a dark colour-inducing neurohormone, a lipophilic neuropeptide (Pener and Yerushalmi 1998). The yellow and black pattern acts as a warning coloration because the locusts start consuming toxic plants when population density increases, thus becoming unpalatable to their predators (Sword *et al*. 2000). Theory predicts that warning coloration is favoured by selection provided that the density of individuals is high enough for frequent predator–prey encounters to facilitate avoidance learning in predators (see Section 2.4.5). Toxicity in combination with warning coloration at high density results in decreased predation and can thus potentially contribute to a further increase in density. Interestingly, the behavioural changes in individuals are reversible if densities are reduced. The longer locusts were under crowded conditions, the longer it took them to return to the solitary state at low densities.

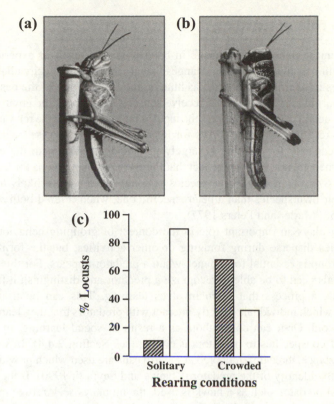

Fig. 8.4   Different phenotypes of desert locust. (a) The green solitary form. (b) The yellow and black gregarious form. (Photos by S. J. Simpson.) (c) Contrast in behaviour of solitary- and crowd-reared locusts when given a choice to move towards a stimulus group of conspecifics (hatched bar) or away from it (open bar). (Adapted from Roessingh *et al.* 1993.)

Females that were raised under low density conditions but kept under high density ones during oviposition, produced gregarious offspring, indicating that they could modify the behavioural tendencies of their offspring according to environmental conditions as late as oviposition. McCaffery *et al.* (1998) isolated a component in the egg foam that is responsible for the control of the offsprings' gregarious behaviour. Removal of the glands that produce this component resulted in largely solitary offspring. This flexibility in behaviour is an astonishing adaptation to extremely changeable environmental conditions. Whether locusts find themselves at high density is driven, among other factors, by the heterogeneity of the resource distribution at small spatial scales. The more aggregated the food resources, the more likely the locusts within a solitary population will end up in a given crowd, against their predisposition to avoid each other. The crowded conditions then trigger the transition to the gregarious form, and thereby change a passive congregation into an active aggregation.

8.6 The role of learning

Learning can be defined as a change in behaviour as a result of experience. This makes learning distinct from other changes that are due to ageing, circadian rhythms, and changes in internal states (such as hunger and fear). Some of the basic requirements of learning are the ability to receive sensory input about the environment via receptors, a neural network to process the input, a memory to store the relevant information, and a recall system that allows for information to be accessed when needed. While the capacity for learning is largely genetically determined, the contents of what is learned are not, despite the fact that there are predispositions for learning certain things over others (e.g. some species of songbirds are more likely to learn the song of their own species than a heterospecific one, when offered both song types; Thorpe 1958; Marler and Peters 1977).

Learning plays an important role in the context of grouping behaviour because many species disperse during foraging or other activities, but the formation of a cohesive group is essential for defence when a predator appears. For this to be possible, animals need to be able to recognize a predator and distinguish it from harmless species, a process that often involves learning. This can be trial-and-error learning in which individuals directly interact with predators that they learn to recognize and avoid. Or it can come about as a result of social learning, in which the response of conspecifics to predators is copied (see Section 2.4.4). In vervet monkeys, for instance, three different predator alarm calls are used which provide information about the identity of the predator (Cheney and Seyfarth 1990). If the alarm call for an aerial predator such as a hawk is used, the monkeys seek refuge in a bush. If the call indicates the approach of a large cat such as a leopard, then they climb into trees; if it is a snake, then they stand on their hind legs to survey the area. Perception of the alarm call is sufficient to trigger the appropriate response and the animal does not need to perceive the predator itself. The correct response to the different alarm calls is learned by juveniles by watching more experienced group members.

In fathead minnows, it has been demonstrated that predator-naïve individuals will show a fright response to chemical stimuli from a pike (a large predatory fish) when in the presence of predator-experienced conspecifics. Furthermore, they will retain the fright response even when subsequently tested on their own (without experienced conspecifics) indicating that they have learned how to recognize a pike (Mathis *et al.* 1996). Such information transfer is not only possible between individuals of the same species, but also between different species (Klump and Shalter 1984; Krause 1993a; Mathis *et al.* 1996). A particularly interesting study was conducted by Chivers and Smith (1994) who introduced water or alarm substance to a tank of fathead minnows and subsequently exposed them to either a pike or a goldfish. Alarm substance is found in the club cells of the epidermis of many fish species and gets released when a predator (such as a pike) ruptures the skin of its prey. After a period of three days, those minnows that had been initially exposed to pike and goldfish in the presence of alarm substance showed a much stronger anti-predator behaviour (e.g. refuge use, shoaling behaviour) to pike and goldfish stimuli than those which had been given just water. The anti-predator response to pike and goldfish was about equally strong. Two months later the test was repeated and fathead minnows now responded more strongly to the pike stimuli than to goldfish ones, suggesting that they have a predisposition for recollecting the cues of their natural predators. The importance of predispositions for learning predator-related cues has also been demonstrated in studies on behavioural development. Experience early in life with predators or predator-like objects can be important for the development of shoaling behaviour and general anti-predators behaviours in fish (Magurran 1990; Huntingford *et al.* 1994). Juvenile European minnows (14 mm long) that came from a pike-sympatric population were split into two groups, only one of which was exposed to a pike model. Two years later the shoaling behaviour of the adult fish was tested in response to an approaching predator model. Model-experienced fish showed a far stronger shoaling response than naïve ones (Magurran 1990). This suggests that the early experience with a predator had an important influence on the behavioural development of the individual. The same procedure was carried out using minnows that came from a pike-allopatric population. No influence of early experiences with predators on shoaling behaviour was found, suggesting that only the pike-sympatric population had a predisposition to respond adaptively to early experiences with predators.

The importance of social learning has so far only been discussed in the context of predator stimuli (e.g. the vervet monkeys and fathead minnows above). However, it has also been much investigated with regards to foraging behaviour (Fig. 8.5). Laland and co-workers demonstrated in a number of elegant experiments how information that has been learned by one individual can be transmitted to others within the same group and also to subsequent generations via cultural transmission (Laland and Reader 1999; Laland and Williams 1998; Reader and Laland 2000). A shoal of four

Fig. 8.5 Observational learning in Japanese snow monkeys. One individual watches another group member wash a potato, a behaviour that was later copied and spread through the troop. (Adapted from Goodenough *et al.* 1993.)

fish was trained to use one of two routes to access food in their tank. After training was achieved, one fish was replaced by a naïve new one every day. The shoal of untrained fish continued to use the same foraging route even after all trained individuals had been replaced, demonstrating the transmission of information between group members (Laland and Williams 1997). In the same way, information can be transmitted from adults to juveniles, allowing vertical transmission of information between generations. Given that many species prefer to group with conspecifics of the same size, that are familiar and parasite free, information transfer via social learning is likely to be non-random within animal populations (Lachlan *et al*. 1998; Krause *et al*. 2000b; see also Chapter 6 on group composition). In both of the above examples (monkeys and fish), the mechanisms of learning involved are based on social learning, where the copying of the behaviour will result in positive or negative consequences (Pearce 1997).

Whether or not an animal learns individually via trial-and-error or by copying the behaviour of group members should in part depend on the environment they live in. In rapidly changing environments, trial-and-error learning should be selected for because information needs to be updated regularly. In contrast, relatively stable environments should select for social learning or potentially for a mix of social learning and trial-and-error learning. What rate of environmental change is required to select for which type of learning remains controversial and should be an interesting field for future research.

8.7 Parasite-mediated changes in behaviour

Behavioural changes in host organisms can either be a result of the host's reaction to the parasite (and therefore be beneficial to the host) or a consequence of host manipulation by the parasite (and thus beneficial to the parasite). The latter is believed to be particularly selected for among parasites with complex life cycles that involve one or more intermediate hosts. According to the 'adaptive manipulation hypothesis', the parasite manipulates the host's behaviour in ways that increase the probability of transmission to the next intermediate or final host of the parasite (Poulin 2000). In some cases (see below), host manipulation involves counteracting behavioural strategies such as grouping that would normally protect a host from its predators (Radabough 1980; Krause and Godin 1994a; Barber and Huntingford 1996).

The life cycle of *Crassiphiala bulboglossa*, a trematode worm, involves the belted kingfisher (*Megaceryle alcyon*), its final avian host, and two intermediate hosts, a freshwater snail and different species of freshwater fish (Fig. 8.6). The worm reproduces in the kingfisher, the eggs leave with the faeces, and the newly hatched miracidium infects a snail. After leaving the snail the cercariae float in the water column until they come into contact with fish, whose body they penetrate to encyst in the musculature as metacerariae, which are externally visible as black spots. Parasitized fish are found more often in front and peripheral positions of shoals (Krause and Godin 1994a). Shoaling behaviour is reduced with increasing

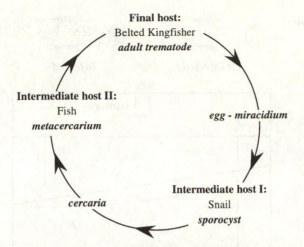

Fig. 8.6 Life cycle of the trematode worm, *Crassiphiala bulboglossa*, which involves a final avian host, the belted kingfisher, and two intermediate hosts, a freshwater snail and different species of freshwater fish.

parasitism and individuals with high parasite loads are often found alone (Ward *et al.* 2002). These behavioural changes are likely to make the fish more vulnerable to its predators (including the belted kingfisher) (Fig. 8.7). Interestingly, the parasitic cysts are particularly common dorsally, which may make the fish visually more conspicuous to aerial predators. Furthermore, parasitized individuals are avoided by conspecifics based on the presence of the black spots, which leads to the segregation of parasitized and unparasitized fish in the wild (Hoare *et al.* 2000). Work by Lafferty and Morris (1996) on a similar host–parasite system showed that the parasite manipulations of the host do indeed result in a higher predation-related mortality of parasitized fish, increasing transmission of the parasite to the final host. The mechanism that brings about the above changes in fish behaviour is not known, but could potentially involve starving the host, which is then forced to leave the group to reduce competition from group mates for food. A reduction in shoaling behaviour similar to that found in the killifish was also observed in minnows that were infected by endoparasites, that have piscivorous birds as their final hosts (Radabough 1980; Barber and Huntingford 1996).

One of the questions that arises from such studies is whether a reduction in the shoaling tendency is really an alteration of host behaviour exclusive to parasites with complex life cycles. One might expect that all parasite species result in the weakening of the host and thereby cause a change in shoaling behaviour. Lemly and Esch (1984) reported, however, that the swimming behaviour of juvenile bluegill sunfish, *Lepomis machrochirus*, was not influenced by parasite infection of a trematode worm, *Uliver ambloplitis*. Similarly, the frequent occurrence of parasitized killifish in front positions of shoals does not indicate that infection automatically results in reduced swimming performance (Ward *et al.* 2002). Another issue concerns the

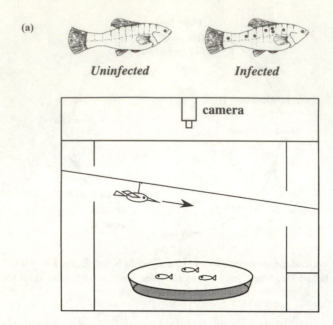

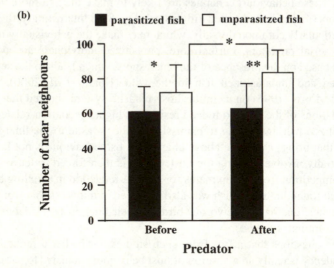

Fig. 8.7 (a) A mixed shoal of parasitized and unparasitized banded killifish was presented with a simulated attack using a stuffed belted kingfisher. (b) Parasitized banded killifish showed a weaker tendency to shoal than unparasitized ones and did not increase their shoaling behaviour after a simulated predator attack (see panel *a* for experimental set-up). (Adapted from Krause and Godin 1994a.) (c) Parasitized killifish suffered a higher predation rate than unparasitized conspecifics. (Adapted from Lafferty and Morris 1996.)

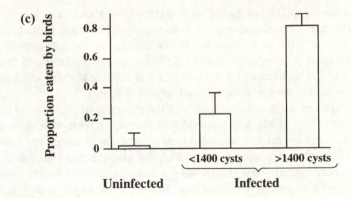

Fig. 8.7 *Cont.*

timing of the manipulation, which should take place when the parasite is ready to be transmitted to the next (or final) host in order to be most effective in aiding parasite transmission. And finally, it is essential to determine how effective a manipulation is in terms of getting the parasite to the desired final host. An abundance of studies demonstrated that parasite infection results in a behavioural change in hosts and in some cases also in higher mortality due to predation (see Barber *et al.* 2000 for a review). However, very few studies also show that behavioural changes in hosts really lead to a higher transmission probability of the parasite to its final host(s) and not just to any predator.

8.8 Summary and conclusions

We started this chapter by looking at some of the ontogenetic constraints of grouping behaviour. In many *r*-selected species of fish, large numbers of eggs are produced leading to planktonic larvae that are not capable of moving against the currents. With the development of red–white musculature, positional control becomes possible, and an increasing differentiation of the lateral line and the eyes facilitate the location of conspecifics in the surroundings. All of which are important prerequisites for the onset of grouping behaviour. In contrast, in *K*-selected organisms, the offspring tend to be relatively larger and more developed at birth (or hatching) and fully capable of moving about with a group of conspecifics.

During ontogeny the plasticity of the individual response to differences in rearing conditions allows for adaptive changes in both morphology and behaviour. In many social insects, the phenotypic development of larvae is largely under hormonal control. The role of the brain and neural stimuli as an interface between hormones and nutritional factors needs further investigation regarding caste determination in social insects. The case study of the desert locust shows that changes in grouping behaviour can take place within just a few hours and have a lasting effect on the next generation.

The detailed understanding of the mechanisms that bring about such changes in social tendencies at the hormonal and neuronal level remains a challenge for future research.

Learning as juveniles or adults is another factor that can mediate grouping behaviour that has been demonstrated in many different species. Learning in this context can be involved in a number of different tasks that include the correct identification of predators and the choice of the best anti-predator strategy.

The final section of this chapter dealt with the influence of parasites on the grouping behaviour of their hosts, which provides an alternative route to behavioural modification and is not necessarily restricted to any particular developmental stage of an individual. Few examples are known in which the adaptive manipulation hypothesis (AMH) has been properly tested, including a demonstration of higher transmission probabilities of parasites to their final host. Further work on the underlying mechanisms of host manipulation is needed to provide a better understanding of these intriguing systems. Among those studies that have examined the AMH, very few have specifically looked at changes in grouping behaviour. Thus there is a lot of scope for future studies.

9

Mechanisms

9.1 Introduction

Grouping behaviour requires the recognition and localization of suitable group mates, which can be either con- or heterospecifics (Section 9.2). Most of the research, however, has centred on the question of how conspecifics are recognized. As we discussed in Chapter 1, groups are brought about by social attraction between individuals that are usually found in close spatial proximity. In this chapter, we take a closer look at what the term 'social attraction' actually means (Section 9.3) and how it can be measured and modelled in terms of inter-individual distance regulation (Sections 9.3 and 9.5). First we will consider group formation under predation threat. This is a relatively simple situation in which only the 'attractive forces' between individuals matter, as they try to form a group as quickly as possible so as to reduce predation risk (Section 9.4). This will be followed by a general discussion of collective behaviours, such as group locomotion and foraging, where individuals often have to balance conflicting demands on spacing that require more complex decisions (Section 9.5). Finally, we will discuss how animals make assessments regarding group size and group composition (Section 9.6) and discuss the need for incorporating these assessment abilities into models of group size distributions (Section 9.7).

9.2 Recognition of suitable group mates

The mechanisms involved in recognizing conspecifics, close kin, or familiar individuals and in remembering specific individual identities have been the object of intense research in recent years. Species recognition received great attention from researchers in the context of mate choice, mainly because of its potential role for the process of speciation. Similarly, the ability to recognize kin, familiars, and/or specific individuals has important implications for the evolution of altruism. The fact that assortative behaviour in animals is so common suggests that many, if not most, species have evolved mechanisms for the recognition of conspecifics, familiar individuals, and in some cases also of kin. However, the identification of the specific cues and sensory channels involved in the recognition process, and the extent to which such recognition is genetically predisposed or learned, have been objects of much dispute. There are three basic prerequisites for active assortative behaviour of any kind:

1. A cue has to be produced which is indicative of phenotypic and/or genotypic trails.
2. The information regarding this cue has to be received and processed.
3. The information is then acted upon (Hepper and Cleland 1998).

   In many species of mammals and fish, recognition mechanisms involve olfactory cues that are associated with the major histocompatibility complex. The olfactory cues involved are usually learned shortly after birth in mammals. Human mothers and their newborn infants are, for instance, capable of recognizing each other based on olfactory cues after breast feeding has started (Porter 1998). Piglet siblings that were separated a few days after birth and then kept apart for several weeks showed a preference for familiar non-sibs over unfamiliar sibs, which indicates that association preference was based on learning (Stookey and Gonyou 1998). Recent work on golden hamsters, however, has suggested that kin recognition in mammals does not necessarily have to involve learning (based on postnatal experiences with close kin) but could be a result of individuals comparing the odour of conspecifics with their own, a process which has been termed 'self-matching' (Mateo and Johnstone 2000). In primates, visual cues have also been shown to be of importance for both species and kin recognition (Fujita *et al.* 1997; Parr and de Waal 1999), and auditory ones have been reported to play a role in seals (Philips and Stirling 2000). Several species of fish use electrical stimuli for identifying conspecifics (Kramer and Kuhn 1994) and in social insects, olfactory cues are the norm for kin recognition, with both genetic and environmental components being involved (Downs and Ratnieks 1999). In the latter, individuals that are isolated from the nest for a period of several days can be reintegrated. However, after a period of several weeks, reintroduction resulted in aggressive responses of nest-mates, suggesting that the odour of the isolated individual had changed over that time (Boulay *et al.* 2000). Nest-mate recognition is usually based on cuticular odours (such as hydrocarbons), whose characteristics are maintained by frequent mandibular interactions between workers exchanging food. If the workers are isolated from the colony for too long, then their cuticular odours change to the degree that they are no longer recognized as nest-mates.

9.3 Inter-individual distance regulation: attraction and repulsion

In many species of group-living birds and mammals, individuals (particularly juveniles) that get separated from their group mates show signs of physical distress that are often accompanied by vocalizations: birds (Gaioni and Ross 1982; Launay *et al.* 1993), small mammals (Carden and Hofer 1990; Hennessy *et al.* 1991; Carbonaro *et al.* 1992). Whether these vocalizations are mainly due to physical discomfort or to social isolation is unresolved. However, in experiments where the ambient

(a) **Latency of the first peep** **(b)** **Number of peeps**

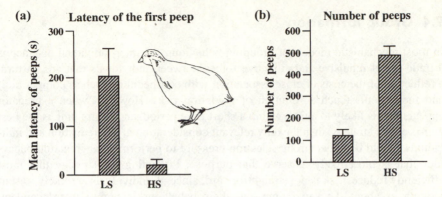

Fig. 9.1 Influence of a 15 minute social isolation period on distress calling in two lines of Japanese quails, *Cortunix japonica*, which have been selected for low (LS) and high levels (HS) of sociality. (a) Mean latency of first peep. (b) Total number of peeps. (Adapted from Launay *et al.* 1993.)

conditions (such as temperature) were kept constant, it could be demonstrated that social isolation alone could trigger vocalization in young birds (Launay *et al.* 1993) (Fig. 9.1). The physiological consequences of social isolation include changes in hormone levels and in the immune system. In mammals, the behavioural response can be quantified in terms of the number and intensity of vocalizations. Motivation to re-join the group can be assessed by making the individual work in a treadmill (Mills and Faure 1990). In fish, the motivation to re-join a shoal can been measured as the time a single individual spends in association with a stimulus shoal in an adjacent tank (Hager and Helfman 1991).

It is evident from this work that group-living individuals are attracted to each other and that there are mechanisms in place to achieve group formation if individuals get separated. On the other hand, it has been observed that social attraction rarely results in actual physical contact between individuals. In most cases, an approach stops when two individuals come within about a body length of each other. Indeed, direct physical contact often results in avoidance behaviour. As a result of such attraction and repulsion, a species-specific inter-individual distance can be observed in most group-living animals from a wide range of taxonomic groups: insects (Kennedy and Crawley 1967; Dicke 1986), fish (Partridge 1980). The underlying function of attraction between individuals is based on the benefits of grouping (Chapter 2). And it has been suggested that the avoidance behaviour (when two animals get too close) has evolved to ensure that animals keep sufficient space between each other to facilitate efficient escape performance. For instance, individuals in a flock of birds that are on the ground would not be able to fly off if they were too close to each other. Very close contact could also interfere with foraging behaviour and increase competition (Chapter 3).

9.4 Group formation

In most mechanistic models of grouping behaviour, the inter-individual distance is a trade-off of repulsive and attractive forces between individuals that are, in turn, a reflection of the costs of close association with conspecifics (such as competition) and the benefits (such as dilution of predation risk). However, when a predator appears that is likely to attack within a short time period, and if grouping is the best anti-predator strategy, then the only relevant consideration is risk reduction and individuals should be under strong selection pressure to perform an aggregation behaviour that is most likely to serve that purpose. Predicting the strategy that most efficiently reduces risk is not straightforward, since all individuals are likely to start moving at about the same time and their behaviour is highly interdependent. Hamilton (1971) in his pioneering Selfish herd paper predicted that each individual should approach its nearest neighbour. However, this hypothesis has been criticized on various grounds. First, the nearest neighbour in space is not necessarily the one that can be reached fastest. It has been demonstrated in small freshwater fish that were frightened with an undirectional light stimulus, that individuals were most likely to approach their nearest neighbour in terms of approach time and not in terms of approach distance (Krause and Tegeder 1994). Secondly, it has been pointed out that the approach of the nearest neighbour would not result in the most efficient reduction in risk possible. Morton *et al.* (1994) demonstrated that higher order movement rules according to which an individual moves between its first and second nearest neighbour would be more advantageous. However, to date no empirical analysis of aggregation behaviour under predation threat has critically tested any of these models.

9.5 Collective behaviour

9.5.1 Locomotion

A challenge to be faced by empirical investigations of the mechanisms of grouping behaviour is that the individual behaviour of group members tends to be highly interdependent. In particular, it is difficult to unravel the role of individual behaviours in decision-making processes (such as directional choice of animal groups) that are characterized by high degrees of synchronization of individuals and very short response latencies. Over the last 10 years, substantial progress in our understanding of the mechanisms of grouping behaviour has been made with the use of individual-based computer simulations (Huth and Wissel 1992; Reuter and Breckling 1994; Gueron *et al.* 1996; Levin 1997; Parrish and Turchin 1997; Flierl *et al.* 1999). Most of these simulations make the assumption that each individual changes its direction (of locomotion) and speed in response to its conspecifics. This response takes the form of attraction to and alignment with conspecifics, or repulsion from them, if they approach too closely (Fig. 9.2) (see Warburton 1997 for a review). Thus individuals

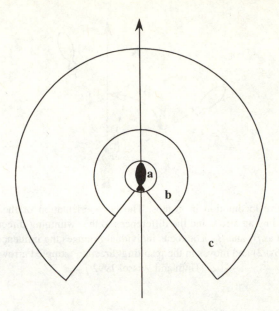

Fig. 9.2 Distance regulation model (adapted from Huth and Wissel 1992). (a) The repulsion zone. (b) The zone within which the individual aligns itself to conspecifics. (c) The area within which an approach towards a conspecific is performed. The direction of locomotion is indicated by an arrow. The area directly behind the fish is a blind zone in which the fish does not respond to conspecifics.

change their direction (of locomotion) and speed in response to surrounding conspecifics. The simulations differ with regards to the more detailed assumptions underlying individual behaviour and the actual parameter settings. Huth and Wissel (1992), for instance, suggested that an individual responds only to its four nearest neighbours; whereas in Reuter and Breckling's (1994) simulation, all group mates are taken into account, although their influence on the focal individual decreases with distance (Fig. 9.3).

The above mentioned simulations demonstrate that a general direction of locomotion of the entire group can arise from local interactions of group members in a self-organized fashion and without leadership by dominant individuals. Furthermore, the models generated grouping behaviour whose cohesion and degree of polarization was largely consistent with that of real animals, although few direct tests of the models have been carried out so far (Parrish and Turchin 1997). Using these simulation models to predict the behaviour of individual group members from one time step to the next should allow a quantitative test of the models and identification of the model that performs best (Fig. 9.3).

Another area that we know little about is the frequency with which individuals update information on their position relative to other shoal mates, and how quickly decisions on position-related changes can be executed. It is to be expected that this is context-dependent with higher update rates when a predation threat is perceived.

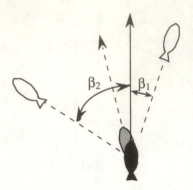

Fig. 9.3 Collective locomotion in groups. The body orientation of the focus individual (black) is indicated by an arrow and the differences to the swimming directions of two conspecifics are given as β_1 and β_2. The focus individual averages the influences of its two near neighbours $(\beta_1 + \beta_2)/2$ and moves in the resulting direction (stippled arrow). (Adapted from Huth and Wissel 1992.)

9.5.2 Positioning behaviour and leadership

Inter-individual distance regulation is under motivational control. Animals decrease their distance to group mates when frightened (Krause and Tegeder 1994) or after feeding (Hunter 1966) and increase it if hungry (Keenleyside 1955; Robinson and Pitcher 1989; Romey 1997). In the latter case, this can ultimately lead to individuals leaving the group and/or the disbanding of the entire group. Small motivational differences between group members, however, provide a promising starting point for investigating differential positioning of individuals within groups.

We discussed in Chapter 5 that the positioning behaviour of individuals depends on their internal state. Hungry individuals are more likely to occupy peripheral group positions where food availability is higher; frightened individuals are more likely to be found in the centre. There are different ways in which we can imagine that position changes come about. An individual could assess the group structure and then position itself relative to that. However, it seems unlikely that such an assessment is always possible, particularly if group sizes are large. Another way for an individual to achieve positional changes is by changing its locomotion parameters (such as speed and turning angles) and spacing behaviour (e.g. inter-individual distance). Food deprivation typically leads to an increase in locomotory activity and an increase in inter-individual distance (Wieser *et al.* 1988). A generic simulation model showed that individuals with a small zone of repulsion tend to be at the front or near the centre of the group whereas variation in the zones of alignment and attraction had little effect on positioning behaviour (Couzin *et al.* in press). High turning rates were negatively correlated with being at the front of the group and positively correlated with centre and rear positions. Increased speed of locomotion was sufficient to explain frontal and peripheral positioning in individuals within groups (Romey 1996; Couzin *et al.* in press), which may explain why hungry whirligig beetles are found more often on the edge of groups (Romey 1995). Further, Gueron *et al.* (1996) demonstrated how differences in speeds of locomotion between group members can produce the phenomenon of leaders and trailers within groups in Grevy's zebra and other ungulates. In contrast, aggressive behaviour of dominant individuals towards subordinates has been shown to result in central positions of high-ranking group members in primates (Hemelrijk 2000). These examples illustrate how very simple local interaction rules with group mates can produce differences in positioning behaviour without any information on the overall group structure being used.

We stated above that directional locomotion is possible without the existence of leadership. However, there is increasing empirical evidence for some group members having a greater influence on the direction of locomotion than others (Bumann and Krause 1993; Reebs 2000). Hungry individuals tend to occupy front and peripheral positions and their greater locomotory activity and inter-individual distance are likely to result in the initiation of new directions of locomotion (compared with well fed individuals that will stay well inside the group). They are also more likely to stick to their direction of locomotion, even at the cost of getting temporarily separated from the group. Reebs (2000, 2001) designed a very elegant experiment that clearly demonstrated the strong influence that small subsets or even single individuals can have on an entire group. A shoal of 12 golden shiners (*Notemigonus crysoleucas*, a finger-long cyprinid fish) were trained to swim to a specific location in a large tank (1.2 × 1.8 m) where they were fed at the same time every day (Fig. 9.4). The tank was divided into two areas, one of which was shaded and preferred by the fish, the other was brightly illuminated and was where the food was delivered. In experiments with shoals of trained and naïve fish, in which the trained fish were always in the minority (down to a single fish in a shoal of 11 naïve ones), it was demonstrated that even a few trained fish were capable of leading an entire shoal of naïve fish into the brighter tank area at the time when food was expected, even if none was provided. In the absence of trained individuals in the group, naïve fish were not found in the feeding area at that time (Fig. 9.4). Although none of the fish were marked, Reebs concluded from the analysis of his video recordings that the trained fish positioned themselves at the front of the shoal.

Groups which are composed of both informed and uninformed individuals are frequently found in the context of annual migration patterns of birds, ungulates, and fish where some of the older individuals have already made the journey at least once before whereas the young of the year are naïve. The modelling of such leadership phenomena provides an interesting challenge for further studies.

9.5.3 Group structure

Changes in the behavioural zones of alignment and attraction (see Fig. 9.2) can have a strong impact on the overall group structure (Couzin *et al.* in press). Low levels of alignment among group members produce what is known as a swarm where individuals perform attraction and repulsion behaviours but show no polarization. This type of group is commonly found in aggregations of mosquitoes and

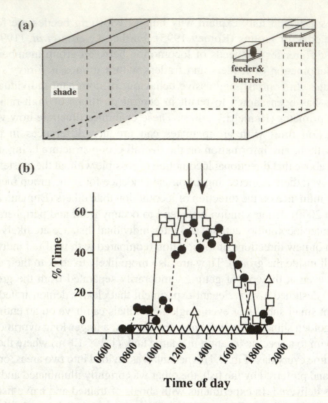

Fig. 9.4 (a) Experimental set-up for an investigation of leadership in fish. The shaded part of the test tank was preferred by the golden shiners but they learned to swim to the opposite side of the tank when food was provided at midday. A floating barrier was used to restrict the food flakes to the corner. (b) Percentage time spent near the feeder by untrained fish (open triangles), trained fish (open squares), and a mixed shoal of three trained and nine untrained individuals (filled circles). (Adapted from Reebs 2000.)

chironomids (Ikawa and Okabe 1997). A so-called torus formation (in which the individuals circle around an empty core) occurs when the zone of alignment is small and the zone of attraction relatively large. Torusses have been observed in a number of pelagic fish species including barracuda, jack, and tuna (Parrish and Edelstein-Keshet 1999). The torus allows fish to remain in the same place while continuing to swim, which is necessary for respiration in several pelagic species. The energetic costs of swimming are potentially reduced because each individual swims in the slipstream of another (see Chapter 2 on benefits of grouping).

More mobile groups are produced if the zones of alignment (between individuals) and attraction are both large. The resulting groups exhibit high degrees of polarization that are characteristic of many species of flocking birds and shoaling fish. In particular an increase in the zone of alignment can produce highly organized groups in which all individuals appear to be synchronized in their behaviour: as has been reported from flying flocks of wading birds (Heppner 1997) and schools of pelagic fish (Pitcher and Parrish 1993).

9.5.4 Collective foraging

Our understanding of group level phenomena such as collective locomotion (see above), phenotype assortedness (Chapter 6), and division of labour has been greatly enhanced by invoking the process of self-organization (Bonabeau *et al.* 1997; Sendova-Franks and Franks 1998). This is a mechanism by which properties of a system emerge from multiple interactions between its components (i.e. usually individuals), whose interaction rules do not directly code for the system properties. A good example of this is superefficient teams in army ants (Franks *et al.* 1999; Anderson and Franks 2001). A team of *Eciton* ants can carry a much greater weight than the sum of the individual efforts would suggest. Army ants form some of the largest known animal groups with up to 20 million members. They perform swarm raids (consisting of tens of thousands or even hundreds of thousands of individuals) on a regular basis to collect prey that can take them several hundred metres away from their nest. The structure of these swarm raids depends largely on the distribution of their prey (Fig. 9.5). High densities of evenly distributed prey result in a broad swarm front; clumped prey distributions break the swarm patterns into sub-swarms.

Communication in army ants is largely based on olfactory cues and each ant leaves a pheromone trail behind it wherever it goes. Individuals have a tendency to follow

Fig. 9.5 The structure of army ant swarms depends on the prey distribution which can be (a) even or (b) clumped. (Adapted from Deneubourg *et al.* 1989.)

the pheromone trails of nest-mates, which reinforces existing trails. Computer simulations of the swarming behaviour of these ants showed that a relatively simple set of rules at the level of the individual ant was sufficient to produce many of the complex behaviours observed at the swarm level during foraging behaviour (Deneubourg *et al.* 1989). Interestingly, army ants are found in both the Old and the New World which means that they have evolved apart for millions of years. Nevertheless the way they behave during swarm raids is remarkably similar, which suggests that they independently converged on similar organizational principles. If, as models of self-organization suggest, many principles of social organization are indeed based on simple rules at the individual level, then the mutations in behaviour required to produce systems of the complexity we observe in many eusocial insects, are fewer and of a less far-reaching nature than previously believed. As Franks (2001) put it: 'Self-organization theory does not suggest that natural selection has had no role in the creation of certain patterns in biology—rather it suggests that natural selection has had rather less to do than one might expect given the complexity of the global structure'.

9.5.5 Teams

Animal teams can be defined as a special form of groups in which there is division of labour with each group member concurrently performing a different subtask. Furthermore, the performance of subtasks is interchangeable between group members (Anderson and Franks 2001). Team behaviour is frequently observed in eusocial insects. However, due to caste differences, interchangeability of subtasks is not always possible as in the following example. When defending their colony, some ant species are known to divide up tasks with smaller workers holding on to an intruder whereas larger individuals are required for killing it (Detrain and Pasteels 1992). Team behaviour has also been reported from mammalian species and in particular those which engage in reciprocal altruism and mutualism such as vampire bats, olive baboons, and group-hunting species such as wild dogs and lions (Stander 1992; Anderson and Franks 2001). In vampire bats usually only a small proportion of the colony members is successful in obtaining a blood meal on a given night and starvation can occur over a relatively short time period of just a few days (Wilkinson 1984). Unsuccessful individuals have been observed to obtain blood from successful colony mates which ensured their survival. The 'favour' of donating blood was later returned, with pairs of bats engaging in reciprocal altruism. In this case, the roles of donor and receiver were exchanged on a regular basis. In group-hunting species such as lions, individuals are known to carry out subtasks in that they take up different spatial positions during a chase (Stander 1992). A similar behaviour has been reported from tuna (Partridge 1982).

9.6 Assessment of group size and group composition

Numerous experiments have demonstrated that many species are capable of distinguishing between groups of conspecifics that differ in size. Most of this work, however, was done using small freshwater fish because of the relative ease with which experiments can be carried out (VanHavre and FitzGerald 1988; Hager and Helfman 1991; Krause and Godin 1994b; Tegeder and Krause 1995; Reebs and Saulnier 1997). Hager and Helfman reported that fathead minnows, *Pimaphales promelas*, could detect a difference between small shoals of two and three fish, but also between larger ones of 18 and 23 individuals. In the presence of a predator, where the motivation for shoal choice would be expected to be high, minnows decided more quickly which shoal to join, and avoided very small shoals more strongly. It has been shown that the time required for such assessments makes up a large proportion of the overall response time in fish, and if pushed below the required assessment time, shoal choice becomes random (Krause *et al.* 1997). One likely factor to be involved in group size assessments is the activity of the group members. Experiments in which the activity of group members was manipulated using water of different temperatures showed that an initial preference for the larger of two groups could be reversed if the larger group was kept in cooler water and therefore exhibited overall less activity than the smaller one (Krause and Godin 1995; Pritchard *et al.* 2001).

Similarly there are several studies demonstrating the ability of individuals to distinguish between groups that only differ in composition but not size (Ranta and Lindström 1990; Ranta *et al.* 1992; Peuhkuri 1999; Ward and Krause 2001). Many species of fish are known to prefer conspecifics of matching body length to smaller and larger ones. However, there is the problem of distinguishing between real and apparent size of objects (Wetterer and Bishop 1985). A large fish far away may seem smaller than a small fish which is close by. One (yet untested) possibility is that fish (and possibly other species as well) use allometric relationships to judge object size. For example, the eye size of most juvenile vertebrates is much larger relative to their body than that of adults. This, however, still leaves the interesting question open of how an animal obtains information about its own size.

9.7 Group size distributions

Group size distributions have been described for a number of different species: African antelopes (Wirtz and Lörscher 1982), fish (Seghers 1981), bison (Lott and Minta 1983), and follow a general pattern of small group sizes being much more frequent than larger ones (Fig. 9.6a). The way in which the frequencies of group sizes decrease with increasing group size can be characterized by an exponential distribution or power law relationship (Okubo 1986; Bonabeau and Dagorn 1995; Bonabeau *et al.* 1999). The process that has been suggested to generate these patterns makes three crucial assumptions (Okubo 1986):

1. Animal groups are open and individuals are free to join and leave at any time.
2. Individuals mix randomly and no associations between individuals exist.
3. Individuals have a tendency to aggregate with conspecifics but aggregation behaviour is independent of group size.

So, how can we understand the generation of the above mentioned distribution patterns in this context? If every individual has a certain probability of leaving a group within a given time unit, then the number of individuals leaving large groups is on average higher compared with smaller ones over the same time period. This process selects against the formation of large groups, which in turn become rare.

However, two of the above three assumptions are problematic. In many species, strong preferences for certain group mates over others have been observed (see Chapter 6). Therefore it seems highly unlikely that individuals mix randomly (assumption 2). Furthermore, group size preferences have been documented in a number of species (see Section 9.6), indicating that the tendency to aggregate is not independent of group size. In fact assumption 3 neglects the fact that individual fitness is often a function of group size (see Chapter 4 on optimal group size). Re-plotting the data from Fig. 9.6a in a way that shows the distribution of

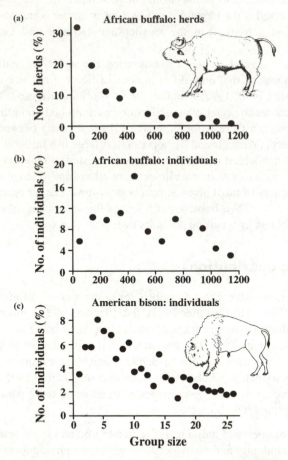

Fig. 9.6 Frequency distribution of (a) group sizes and (b) individuals in African buffalo, *Syncercus caffer* (Sinclair 1977). (c) Frequency distribution of individuals in American bison, *Bison bison* (Lott and Minta 1983).

individuals across groups of different sizes, indicates that the peak of the distributions is found for groups of intermediate size (Fig. 9.6b). Therefore what appears to be a characteristic pattern at the group level is unrepresentative at the individual level. This result highlights the problem of giving an over-proportionately high weighting to small group sizes when these represent only a fraction of the population. Moreover, these small groups could well be special cases of diseased or particularly old individuals.

The distributions of the individuals show that the group size in which most animals were found was considerably larger for buffalo than for bison (Fig. 9.6). It is tempting to conclude that the observed group sizes are also the preferred ones by the animals and that buffaloes therefore form larger groups than bison. However, the group size distributions may be constrained by population density if the latter is low. In the case of the buffalo, the population consisted of about 52 000 individuals in an area of roughly 10 000 km^2, whereas the bison population comprised only 400 on an island of 200 km^2 of which about 86% could be accessed by the animals. Therefore the two species not only differed dramatically in the overall population sizes but also in their densities: 5.2 buffalos/km^2 compared with 2.3 bison/km^2. Therefore when making species comparisons regarding commonly observed group sizes we have to adjust for differences between population densities (and individual and group range sizes). This has not been done in most studies on grouping animals and requires, among other things, knowledge about the functional response between median group size and population density. Median group size should initially increase with population density (because of an increase in encounter frequency) until the preferred group size is reached. Beyond that point, a further increase in population density should mainly result in an increase in the number of groups but not in median group size (Fig. 9.7). The density-dependence of group size and group number is a

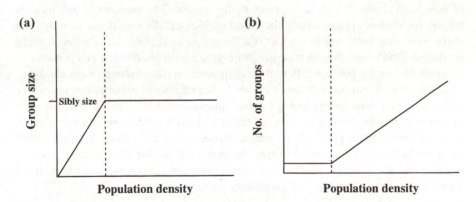

Fig. 9.7 Expected relationship between population density and (a) group size and (b) number of groups. With increasing population density, group size should increase until groups reach the 'Sibly' size (which is larger than the optimal group size; see Chapter 4). At this density, group size should reach a plateau and a further increase in density should result primarily in an increase in the number of groups.

promising area for both further theoretical and empirical work. In particular, manipulation experiments in which the density is changed and the resulting group size distribution observed would greatly increase our current knowledge of the dynamics of grouping behaviour.

9.8 Summary and conclusions

To be able to form groups, animals have to be able to recognize suitable group mates. This can involve a number of different sensory channels, largely depending on the taxonomic group in question. Once a group has formed, the inter-individual distance between group members is the result of attraction and repulsion between individuals. The behaviour of group members is partly based on interactions with other group mates. However, the degree to which grouping individuals use local, as opposed to group level, information remains a matter of dispute. Empirical tests of the existing models of the decision-making processes within groups are very much needed.

Models of self-organization provide one of the most interesting developments of recent years. Powerful computers and sophisticated software have made it possible to study the interactions of virtual animals and the emergent patterns at the level of the group or population. This approach has led to the discovery of surprisingly simple behavioural rules that are sufficient for organizing an entire group or colony to perform behaviours of astonishing complexity and to flexibly respond to changes in the environment.

One of the more controversial areas of individual-based modelling (of grouping behaviour) is the trend to modify individual behaviour such that it produces the phenomena of interest at the group or population level. Thus the behaviour of individual animals is deduced from the characteristics of groups or populations regardless of how individuals of a certain species really behave. The problem is that models taking this kind of approach could be based on biologically unrealistic assumptions at the individual level despite the fact that the group level phenomena are accurately produced. This is possible because the same group level phenomena can potentially be generated on the basis of different local interaction rules between individuals.

A similar problem occurs when it comes to the number of assumptions necessary at the individual level for producing a certain phenomenon at the group level. Models of self-organization have been very useful in identifying the minimum set of rules required for a certain group phenomenon. However, in some cases it seems rather naïve to then claim that real animals are restricted to that minimum of options because this would be ignoring the fact that an organism has been selected to perform a multitude of tasks and has an evolutionary past.

10

Conclusions

10.1 Introduction

In the final chapter of the book we selectively highlight several areas that might benefit from further research. Therefore, this chapter is only to a limited extent a general summary of the book. The emphasis is on the identification of unresolved problems. In particular, we point out where new and promising methodological approaches are available (Sections 10.2.1 and 10.3), and where the discussion would benefit from linking topics that up to now have been treated largely in isolation (Section 10.2.2).

10.2 Group size

Many of the mechanisms underlying the costs and benefits of grouping are fairly well understood (as is evident from Chapters 2 and 3). However, it remains a substantial challenge to integrate the available information to make the predictions from optimal group size models testable. One of the problems in this context is that researchers have focussed on large vertebrates (such as lions) with complex social systems and a long lifespan for which it is very difficult to disentangle benefits and costs clearly and to quantify them accurately. Other study systems (such as colonial spiders or insects) which have a relatively short lifespan, so that actual fitness can be measured (or at least estimated) as a function of group size, might be far more suitable and offer promising avenues for future research. Invertebrates also facilitate the execution and replication of controlled experiments to test specific predictions of optimal group size models, and ethical considerations regarding predation experiments may be less restricting.

Most models of group choice consider only the scenario of a single individual trying to join a group whose members may not necessarily benefit. A frequently found argument is that the newcomer will generally be successful at joining the group because it stands to gain more than the average group member stands to lose. However, these kinds of considerations might greatly benefit from modelling the entire social environment in which the presence of multiple groups is taken into account. After all, a group only has to try harder to repel the newcomer than another group nearby to be successful, which highlights the interdependence of these processes. Whether an individual will stay in a group to which it has gained access should depend partly on whether there are any benefits to be gained from long-term membership. In male hyena, a newcomer usually starts at the bottom of the hierarchy and only gradually climbs to the top, a prerequisite for access to females. Thus the benefits of grouping increase with the time that has been spent in a particular group, which selects against short-term switching between groups. In many species of birds and fish, groups are only formed outside the reproductive season and individual recognition is often difficult, if not impossible, because of large group sizes. Therefore group fidelity is often very low and individuals frequently switch between groups in response to internal state changes and external factors such as a predation pressure.

10.2.1 Individual-based models of group size

Conventional models of optimal group size tend to take a very static view of grouping behaviour, specifying one particular group size which, under certain conditions, should be generally adopted by all individuals in a population. Given that animals undergo ontogenetic and state changes it seems clear that there should be considerable variation in the preferred group sizes between individuals in a population. This is an area in which both theoretical and empirical work is much needed. Individual-based models in which interactions between individuals with different group size preferences are simulated might be a good way forward to generate testable predictions. Individual-based modelling has become a major tool in recent years and has enormous potential for providing insights into highly dynamic social systems where other analytical methods fail. However, model development requires close collaboration between theoretician and empiricist to ensure that the model is based on realistic assumptions before testing model predictions against empirical data. Factors such as the inclusion and specification of attraction and repulsion zones in models of grouping behaviour and the existence and strength of group size preferences, among other concerns, will have to be considered.

For the inclusion of group size preferences into individual-based models, we need more information about the mechanisms of group size assessments. This is a rather neglected but nevertheless important area of research. It has been demonstrated that some species of small freshwater fish are capable of discriminating between group sizes usually involving fewer than 20 members each (Fig. 10.1). It has been suggested that these group size assessments are based on the activity levels of groups (Pritchard *et al.* 2001). However, in the case of large flocks of birds, marine schools of fish, and swarms of krill, which can comprise hundreds of thousands of animals, an individual group member is unlikely to make an assessment of the entire group. In such cases, other mechanisms for group size assessments based on local cues could be at work. It has been reported from divers that some marine fish species form schools that are so dense that the centre of the school receives almost no light. Similarly one would expect the oxygen concentration to be affected by such dense packing of so many individuals. Therefore the local oxygen concentration or light intensity could be an individual's means of assessing group size.

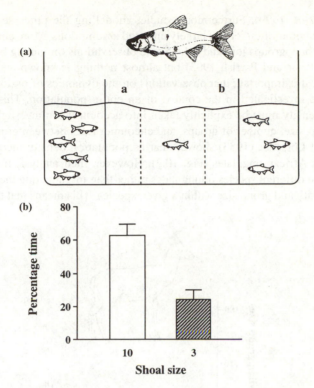

Fig. 10.1 Shoal size preference in golden shiner. (a) Each test fish was presented with a choice between two stimulus shoals of different size (10 fish versus three fish). Time spent in zones **a** and **b** (indicated by dashed lines) was measured. (b) Percentage time spent near each of the stimulus shoals by the test fish. (Adapted from Reebs and Saulnier 1997.)

10.2.2 Population density and group size

Surprisingly few studies have collected information on the range and frequency of group sizes under natural conditions. The available data are mainly restricted to fish: spottail shiner (Seghers 1981), tuna (Bonabeau and Dagorn 1995), and large mammals, e.g. giraffe (Foster and Dagg 1972), buffalo (Sinclair 1977), seven antelope species (Wirtz and Lörscher 1983), American bison (Lott and Minta 1983). These studies clearly show that group size ranges are considerable for most species and can comprise single individuals as well as groups with hundreds or even thousands of members. Our understanding of the processes generating this variation is far from complete and currently even some of the basic assumptions of models attempting to conceptualize this area of research are controversial. Recent attempts at modelling naturally occurring group size frequencies are hampered by the fact that the data that were used came from the fisheries industry which almost certainly underestimates small group sizes (which are of no interest to the fisheries community)

(Bonabeau *et al.* 1999). Furthermore, studies modelling the processes underlying grouping behaviour often rely on largely untested assumptions. More empirical work is needed on how groups join and split. Some observations on joining behaviour are available (Pitcher and Parrish 1993) but almost nothing is known about splitting. Furthermore, it is important that observations on the dynamics of fission and fusion processes are investigated in the context of an entire population. This means that population density must be explicitly taken into account when investigating factors such as group size, number of groups, and encounter rates between groups. A study by Wirtz and Lörscher (1983) showed that as population density increased so did group size in African ungulates (Fig. 10.2). However, their analysis made comparisons between different species of ungulates rather than investigating the relationship between density and group size within a given species. This means that their analysis

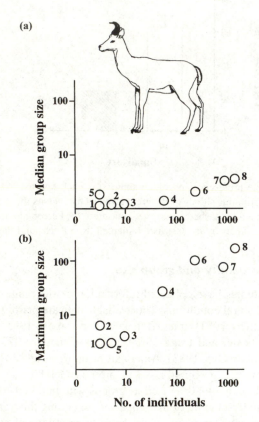

Fig. 10.2 (a) Median group size and (b) maximum group size are plotted against the average number of individuals observed on the transect for each of eight different species of African ungulates: 1, steenbock; 2, Kirk's dikdik; 3, bushbuck; 4, bohor reedbuck; 5, mountain reedbuck; 6, Thomson's gazelle; 7, impala; 8, waterbuck. Spearman rank correlation: median group size $r_s = 0.82$, $p < 0.01$; maximum group size $r_s = 0.86$, $p < 0.01$. (Adapted from Wirtz and Lörscher 1983.)

is likely to be confounded by differences in species ecology and life history factors. In fact their data show that the smaller antelopes like the dikdik exhibit little difference in median and maximum group sizes whereas the larger ones do (see difference in slopes of median and maximum group sizes in Fig. 10.2) which may be explained by the fact that dikdiks are highly territorial. Nevertheless Wirtz and Lörscher's study is a good example of the type of work that is needed.

The positive relationship between population density and group size highlights a problem with many studies on group size that neglect to collect information on differences in population densities between populations or species. The group size differences that were reported could be confounded by differences in population density, which is problematic for many comparative studies of group size (Crook 1964; Jarman 1974; Brashares *et al.* 2000). The population densities of many African ungulate species, for instance, are known to fluctuate enormously between dry and wet years, which could have a strong influence on the grouping patterns exhibited.

In conclusion, Wirtz and Lörscher (1983) showed that group size is positively correlated with population density, but their data do not account for differences in species ecology and life history traits. Brashares *et al.*'s (2000) re-analysis of Jarman's work demonstrated how species ecology and life history traits influence group size, but their data are potentially confounded by differences in population density between species. What is needed is a study that takes the effects of species ecology, life history traits, and population density into account when investigating group size.

An understanding of the dynamics underlying group size distributions also has relevance for conservation biology. Once we know how population density translates into a group size distribution we can make predictions about group size-related mortality rates due to predation. Knowledge of the group size distribution in connection with information on encounter rates between groups could also be important for population management regarding the spread of infectious diseases and for a prediction of information transfer via social learning. For instance, the survival of hatchery-reared animals in the wild could be greatly improved if we gain a better understanding of the role of social learning in groups concerning the optimal ratio of 'tutors' to 'naïve' individuals (in groups) and the optimal group size for learning processes. Finally, in the fisheries industry an understanding of the relationship between population density and group size and group composition might be useful to predict changes in the population density of commercially valuable species from encounter rates with fish shoals of different sizes.

10.3 Comparative studies

Comparative studies have played an important role in our understanding of the ecological factors that act as constraints on grouping behaviour and the life history traits that predispose species towards grouping (Arnold and Owen 1998, 1999).

The recent re-analysis of Jarman's (1974) comparative study of the social organization of African ungulates by Brashares *et al.* (2000) is a good example of how old data sets can be revisited with great benefit. The comparative method has undergone tremendous conceptual changes since Jarman's paper, which contained no formal statistical analysis. The availability of more accurate phylogenies (based on molecular data) for various taxonomic groups, in connection with greater computing power, make re-analysis of some of the other early studies, such as Crook's (1964) detailed investigation of weaver birds, desirable.

A potential problem with comparative studies is that they do not allow determination of cause and effect but merely give correlational evidence. However, once it is known which factors are closely linked, further detailed investigation can provide the information necessary to unravel the underlying mechanism (Arnold and Owen 1998).

10.4 Evolution of grouping

The manipulation experiments in which guppies were transferred from sites of high predation risk to ones of low predation risk probably provide some of the best evidence for evolution in progress because they demonstrated heritable changes in grouping behaviour in an animal population over time (Magurran *et al.* 1995). In general, a number of small tropical species of freshwater fish provide good study systems for investigating evolutionary change in grouping behaviour because (unlike many other vertebrate species) they often produce four to five generations per year, which seems to be sufficient to observe behavioural change over a period of a few years. This is exemplified by Ruzzante and Doyle's (1993) selective breeding experiments using Japanese medaka that showed strong behavioural changes over just two generations. Once the zebrafish genome is fully mapped, this species, which is easy to breed in the laboratory and produces several generations per year, might become an increasingly attractive candidate for evolutionary studies of grouping.

10.5 Group composition

In contrast to the group size issue where much theoretical modelling has been done but not enough empirical work, studies on group composition lack a rigorous theoretical grounding. The basis on which group mates are chosen needs further exploration by modellers (see Ranta *et al.* 1994; Conradt and Roper 2001) as well as empirical testing. Ranta *et al.* (1994) concluded that both movement costs between groups and population density are of particular relevance for phenotypic assortment. Basically, this involves the same factors as we discussed above concerning group size, which is why future modelling should concentrate on developing a unified conceptual framework that incorporates both group size and group composition. The mechanisms by which assortment comes about have often been classified into active choice behaviours and passive sorting. For instance, body length assortment can come about by individuals choosing to be in groups with similar sized conspecifics. Alternatively they always join the groups they encounter but will get 'passively' sorted on the basis of differential speeds of locomotion. Active choice is the more complicated explanation (because it requires additional assumptions about the cognitive abilities of the animal) and should only be invoked if necessary or true. However, it is becoming clear that active and passive mechanisms cannot always be clearly distinguished. If larger individuals move at a higher speed and leave smaller and slower group members behind, then this could also be considered as 'a sort of choice'. Activity synchronization is another factor that might play an important role for group composition, potentially leading to segregation of animals by sex, age, or size (Conradt and Roper 2001). Synchronizing activities is assumed to be costly for group members because it may require the postponement of activities. Conradt and Roper (2001) suggested that mixed phenotype groups are more likely to break up than homogeneous ones because of a conflict of interest between group members. This mechanism could eventually bring about a significant degree of segregation without invoking any preferences for particular group compositions on behalf of the individuals involved.

10.6 Signalling

Another rather underexplored area of research on grouping is signalling. It is well established that certain phenotypic characteristics are preferred in potential group mates (see Chapter 6 on group composition), such as low parasite load and low competitiveness. In some cases it is also known which cues are used by animals to assess these factors in conspecifics. Little information is available, however, on the ability of animals to 'fake' desirable characteristics to gain access to groups. This behaviour should be particularly selected for when the benefits from joining a group are high but where the existing group members can easily expel an unsuitable newcomer. So far this scenario is only known for the heterospecific case where parasites of social insects have evolved ways of passing for a colony member (Hölldobler and Wilson 1990). A useful tool in this context could be playbacks of video sequences which can be edited to decouple morphological and behavioural traits for testing which signals are used in social animals for the assessment of conspecifics (D'Eath 1998) (Fig. 10.3). Alternatively, computer animation can be used to develop virtual animals that, once generated, offer greater flexibility for behavioural manipulation (Kunzler and Bakker 1998). Some of the problems of video playbacks, such as artificiality of colours, two-dimensionality of the stimulus presentation, low image resolution, and lack of interaction between stimulus and test subject, are becoming increasingly solvable with the arrival of more powerful computer technology and imaging software.

10.7 Short-term behavioural change

It has been known for some time that the grouping tendency of animals can change over a very short time span in many species. Our understanding of some of the

Fig. 10.3 A single animal is placed in the central compartment and its response to an image of a conspecific on a monitor is tested. This can be extended to include choice tests between two monitors displaying images of conspecifics that differ in morphology or behaviour. Natural responses to video and/or computer-generated images of conspecifics have been demonstrated in a number of species including primates (Rhesus monkey, Bonnett macaque), birds (domestic chicken, pigeon, zebrafinch), jumping spiders, anolis lizards, and fish (threespine stickleback, guppy).

processes underlying such short-term changes has made great progress with recent work on locusts, which identified key factors of the mechanism underlying the change from solitary to gregarious behaviour. Similar progress has been made with regards to investigating the flexibility in the division of labour in honey bee colonies, which is largely under hormonal control. Other areas of interest include the mechanisms underlying the state-dependent control of inter-individual distances between group members in many species of grouping vertebrates. For instance, food deprivation can lead to an individual changing its spatial position within a group (e.g. back to front position) or even for that individual to leave the group altogether. Results from computer simulations suggest that one possible way of bringing about such changes in behaviour is by increasing the inter-individual distance with increasing hunger level in animals. There is some evidence that groups of hungry individuals are less densely spaced than those of well-fed ones (Robinson and Pitcher 1989). The exact nature of the relationship between nutritional state and inter-individual distance regulation, however, remains unknown.

10.8 Concluding remarks

Writing this book has been much more of a learning experience than we ever expected. We have learnt a great deal about aspects of grouping that we would never have taken the time to explore before. However, no matter where our paper-chases led, we always found vibrant science going on and many interesting questions that have still to be addressed. Our hope is that we have managed to convey this to the reader. We believe that our understanding of the nature of grouping in animals can be greatly improved over the next decade or so, with the new techniques that have become available to us. Why not join us in this research endeavour, and help make this book out of date!

References

Abrahams, M. V. and Colgan, P. W. (1985). Risk of predation, hydrodynamic efficiency and their influence on school structure. *Environmental Biology of Fishes*, **13**, 195–202.

Abrahams, M. V. and Colgan, P. W. (1987). Fish schools and their hydrodynamic function: a reanalysis. *Environmental Biology of Fishes*, **20**, 79–80.

Aksnes, D. L. and Utne, A. C. W. (1997). A revised model of visual range in fish. *Sarsia*, **82**, 137–47.

Alexander, R. D. (1974). The evolution of social behavior. *Annual Reviews in Ecology and Systematics*, **5**, 325–83.

Allan, J. R. and Pitcher, T. J. (1986). Species segregation during predator evasion in cyprinid fish shoals. *Freshwater Ecology*, **16**, 653–9.

Ancel, A., Visser, H., Handrich, Y., Masman, D., and Le Maho, Y. (1997). Energy saving in huddling penguins. *Nature*, **385**, 304–5.

Anderson, C. and Franks, N. R. (2001). Teams in animal societies. *Behavioral Ecology*, **12**, 534–40.

Anderson, D. J. and Hodum, P. J. (1993). Predation behavior favors clumped nesting in an oceanic seabird. *Ecology*, **74**, 2462–4.

Andersson, M. (1976). Predation and kleptoparasitism by skuas in a Shetland seabird colony. *Ibis*, **118**, 208–17.

Andersson, M. and Wiklund, C. G. (1978). Clumping versus spacing out: experiments on nest predation in fieldfares (*Turdus pilaris*). *Animal Behaviour*, **26**, 1207–12.

Andren, H. (1991). Predation: an overrated factor for over-dispersion of birds' nests? *Animal Behaviour*, **41**, 1207–12.

Andrews, R. V. and Belknap, R. W. (1986). Bioenergetic benefits of huddling by deer mice (*Peromyscus maniculatus*). *Comparative Biochemistry and Physiology A*, **85**, 775–8.

Aoki, S. (1982). Soldiers and altruistic dispersal in aphids. In *The biology of social insects* (eds. M. D. Breed, C. D. Michener, and H. E. Evans), pp. 154–8. Westview Press, Boulder, CO.

Armitage, K. B. (1999). Evolution of sociality in marmots. *Journal of Mammalogy*, **80**, 1–10.

Armstrong, E. A. and Whitehouse, H. L. K. (1977). Behavioural adaptations of the wren. *Biological Reviews*, **52**, 235–94.

Arnold, K. E. and Owens, I. P. F. (1998). Cooperative breeding in birds: a comparative test of the life history hypothesis. *Proceedings of the Royal Society London Series B*, **265**, 739–45.

Arnold, K. E. and Owens, I. P. F. (1999). Cooperative breeding in birds: the role of ecology. *Behavioral Ecology*, **10**, 465–71.

Arnold, S. J. and Bennett, A. F. (1984). Behavioural variation in natural populations. III. Anti-predator displays in the garter snake *Thalamnus radix*. *Animal Behaviour*, **32**, 1108–18.

Arnqvist, G. and Byström, P. (1991). Disruptive selection on prey group size: a case for parasitoids? *American Naturalist*, **137**, 268–73.

Avilés, L. and Gelsey, G. (1998). Natal dispersal and demography of a subsocial *Anelosimus* species and its implications for the evolution in spiders. *Canadian Journal of Zoology*, **76**, 2137–47.

Avilés, L. and Tufiño, P. (1998). Colony size and individual fitness in the social spider, *Anelosimus eximus*. *American Naturalist*, **152**, 403–18.

Axelrod, R. and Hamilton, W. D. (1981). The evolution of cooperation. *Science*, **211**, 1390–6.

Badgerow, J. P. (1988). An analysis of function in the formation flight of Canada geese. *Auk*, **105**, 749–55.

Badgerow, J. P. and Hainsworth, F. R. (1981). Energy savings through formation flight? A re-examination of the vee formation. *Journal of Theoretical Biology*, **93**, 41–52.

Baird, R. W. and Dill, L. M. (1995). Ecological and social determinants of group size in transient killer whales. *Behavioral Ecology*, **7**, 408–16.

Baird, T. A. (1983). Influence of social and predatory stimuli on the air-breathing behaviour of the African clawed frog, *Xenopus laevis*. *Copeia*, **1983**, 411–20.

Balmford, A. P. (1991). Mate choice on leks. *Trends in Ecology and Evolution*, **6**, 87–92.

Balmford, A. P. and Turyaho, M. (1992). Predation risk and lek-breeding in Uganda kob. *Animal Behaviour*, **44**, 117–27.

Barber, I. and Huntingford, F. A. (1996). Parasite infection alters schooling behaviour: Deviant positioning of helminth-infected minnows in conspecific groups. *Proceedings of the Royal Society London Series B*, **263**, 1095–102.

Barber, I. and Ruxton, G. D. (2000). The importance of stable schooling: do familiar stickle-backs stick together? *Proceedings of the Royal Society of London Series B*, **267**, 151–5.

Barber, I. and Wright, H. A. (2001). How strong are familiarity preferences in shoaling fish? *Animal Behaviour*, **61**, 975–9.

Barber, I., Downey, L. C., and Braithwaite, V. A. (1998). Parasitism, oddity and the mechanism of shoal choice. *Journal of Fish Biology*, **53**, 1365–8.

Barber, J., Hoare, D. J., and Krause, J. (2000). Effects of on fish behaviour: a review and evolutionary perspective. *Reviews in Fish Biology and Fisheries*, **10**, 131–65.

Barnard, C. J., Thompson, D. B. A., and Stephens, H. (1982). Time budgets, feeding efficiency and flock dynamics in mixed species flocks of lapwings, golden plovers and gulls. *Behaviour*, **80**, 44–69.

Barta, Z., Flynn, R., and Giraldeau, L. A. (1997). Geometry for a selfish foraging group: genetic algorithm approach. *Proceedings of the Royal Society London Series B*, **264**, 1233–8.

Barta, Z. and Giraldeau, L.-A. (2001). Breeding colonies as information centres: a reappraisal of information-based hypotheses using the producer-scrounger game. *Behavioral Ecology*, **12**, 121–7.

Bazin, R. C. and MacArthur, R. A. (1992). Thermal benefits of huddling in the muskrat (*Ondatra-zibethicus*). *Journal of Mammalogy*, **73**, 559–64.

Beauchamp, G. (1999). The evolution of communal roosting in birds: origin and secondary losses. *Behavioral Ecology*, **10**, 675–87.

Bednarz, J. C. (1988). Cooperative hunting in Harris' Hawks (*Parabuteo unicinctus*). *Science*, **239**, 1525–7.

Bednekoff, P. A. and Lima, S. L. (1998). Re-examining safety in numbers: interactions between risk dilution and collective detection depend upon predator targeting behaviour. *Proceedings of the Royal Society London Series B*, **265**, 2021–6.

Bergstrom, C. T. and Lachmann, M. (2001). Alarm calls as costly signals of anti-predator vigilance: the watchful babbler game. *Animal Behaviour*, **61**, 535–43.

Berteaux, D., Bergeron, J. M., Thomas, D. W., and Lapierre, H. (1996). Solitude versus gregariousness: do physical benefits drive the choice in overwintering meadow voles? *Oikos*, **76**, 330–6.

Bertram, B. C. R. (1975). Social factors influencing reproduction in wild lions. *Journal of Zoology*, **177**, 463–82.

Benkman, C. W. (1988). Flock size, food dispersion and the feeding behaviour of crossbills. *Behavioral Ecology and Sociobiology*, **23**, 167–75.

Beukema, J. J. (1968). Predation by the three-spine stickleback (*Gasterosteus aculeatus* L.): the influence of hunger and experience. *Behaviour*, **31**, 1–126.

Bill, R. G. and Herrnkind, W. F. (1976). Drag reduction by formation movement in spiny lobsters. *Science*, **193**, 1146–8.

Black, J. M., Carbone, C., Wells, R. L., and Owen, M. (1992). Foraging dynamics in goose flocks: the costs of living on the edge. *Animal Behaviour*, **44**, 41–50.

Boesch, C. (1994). Cooperative hunting in wild chimpanzees. *Animal Behaviour*, **48**, 653–67.

Bogliani, G., Sergio, F., and Tavecchia, G. (1999). Woodpigeons nesting in association with hobby falcons: advantages and choice rules. *Animal Behaviour*, **57**, 125–31.

Boix-Hinzen, C. and Lovegrove, B. G. (1998). Circadian metabolic and thermoregulatory patterns of red-billed woodhoopoes (*Phoeniculus purpureus*): the influence of huddling. *Journal of Zoology*, **244**, 33–41.

Bonabeau, E. and Dagorn, L. (1995). Possible universality in the size distribution of fish schools. *Physical Review E*, **51**, R5220–3.

Bonabeau, E., Theraulaz, G., Deneubourg, J.-L., Aron, S., and Camazine, S. (1997). Self-organisation in social insects. *Trends in Ecology and Evolution*, **12**, 188–92.

Bonabeau, E., Dagorn, L., and Freon, P. (1999). Scaling in animal group-size distributions. *Proceedings of the National Academy of Sciences of the USA*, **96**, 4472–7.

Bone, Q. and Trueman, E. R. (1983). Jet propulsion in salps (Tunicata: Thaliacea). *Journal of Zoology*, **201**, 481–506.

Boulay, R., Hefetz, A., Soroker, V., and Lenoir, A. (2000). *Camponotus fellah* colony integration: worker individuality necessitates frequent hydrocarbon exchanges. *Animal Behaviour*, **59**, 1127–33.

Bradbury, J. W., Vehrencamp, S. L., and Gibson, R. (1985). Leks and the unanimity of female choice. In *Evolution: essays in honour of John Maynard Smith* (eds. P. J. Greenwood, P. H. Harvey, and M. Slatkin), pp. 301–20. Cambridge University Press, Cambridge.

Bradley, B. J. (1999). Levels of selection, altruism, and primate behavior. *Quarterly Reviews of Biology*, **74**, 171–94.

Brashares, J. S., Garland, T. Jr., and Arcese, P. (2000). Phylogenetic analysis of coadaptation in behavior, diet, and body size in the African antelope. *Behavioral Ecology*, **11**, 452–63.

Breden, F., Scott, M. A., and Michel, E. (1987). Genetic differentiation for anti-predator behaviour in the Trinidad guppy, *Poecilia reticulata*. *Animal Behaviour*, **35**, 618–20.

Briggs, S. E., Godin, J.-G., and Dugatkin, L. A. (1996). Mate-choice copying under predation risk in the Trinidadian guppy (*Poecilla reticulata*). *Behavioral Ecology*, **7**, 151–7.

Brönmark, C. and Miner, J. G. (1992). Predator-induced phenotypical change in body morphology in crucian carp. *Science*, **258**, 1348–50.

Brönmark, C., Pettersson, L. B., and Nilsson, P. A. (1999). Predator-induced defense in crucian carp. In *The ecology and evolution of inducible defenses* (eds. R. Tollrian and D. C. Harvell), pp. 203–17. Princeton University Press, Princeton.

Broom, M. and Ruxton, G. D. (1997). Evolutionarily stable stealing: game theory applied to kleptoparasitism. *Behavioral Ecology*, **9**, 397–403.

Brown, C. R. (1986). Cliff swallow colonies as information centres. *Science*, **234**, 83–5.

Brown, J. L. (1987). *Helping and communal breeding in birds*. Princeton University Press, Princeton.

Brown, C. R. and Brown, M. B. (1986). Ectoparasitism as a cost of coloniality in cliff swallows (*Hirundo phyyhonota*). *Ecology*, **67**, 1206–18.

Brown, G. E. and Brown, J. A. (1992). Do rainbow trout and Atlantic salmon discriminate kin? *Canadian Journal of Zoology*, **70**, 1636–40.

Brown, G. E. and Brown, J. A. (1996). Kin discrimination is salmonids. *Reviews in Fish Biology and Fisheries*, **6**, 201–19.

Brown, J. A. and Colgan, P. W. (1986). Individual and species recognition in centrachid fishes: evidence and hypotheses. *Behavioral Ecology and Sociobiology*, **19**, 373–9.

Brown, C. and Warburton, K. (1999). Social mechanisms enhance escape responses in shoals of rainbow fish, *Melantaenia duboulayi*. *Environmental Biology of Fishes*, **56**, 455–9.

Brown, C. R., Brown, M. B., and Shaffer, M. L. (1991). Food sharing signals among socially foraging cliff swallows. *Animal Behaviour*, **42**, 551–64.

Brown, G. E., Brown, J. A., and Crosbie, A. M. (1993). Phenotype matching in juvenile rainbow trout. *Animal Behaviour*, **46**, 1223–5.

Brown, C. R., Brown, M. B., and Danchin, E. (2000). Breeding habitat selection in cliff swallows: the effect of conspecific reproductive success on colony choice. *Journal of Animal Ecology*, **69**, 133–42.

Bumann, D. and Krause, J. (1993). Front individuals lead in shoals of three-spined sticklebacks (*Gasterosteus aculeatus*) and juvenile roach (*Rutilus rutilus*). *Behaviour*, **125**, 189–98.

Bumann, D., Krause, J., and Rubenstein, D. I. (1997). Mortality risk of spatial positions in animal groups: the danger of being in the front. *Behaviour*, **134**, 1063–76.

Butler, M. J. IV. (1999). The cause and consequence of ontogenetic changes in social aggregation in New Zealand spiny lobsters. *Marine Ecology Progress Series*, **188**, 179–91.

Butler, M. J. IV, Herrnkind, W. F., and Hunt, J. H. (1997). Factors affecting the recruitment of juvenile Caribbean lobsters dwelling in macroalgae. *Bulletin of Marine Science*, **61**, 3–19.

Caine, N. G., Addington, R. L., and Windfelder, T. L. (1995). Factors affecting the rates of food calls given by red-bellied tamarins. *Animal Behaviour*, **50**, 53–60.

Caldwell, G. S. (1981). Attraction of tropical mixed species heron flocks: proximate mechanisms and consequences. *Behavioral Ecology and Sociobiology*, **8**, 99–103.

Canals, M., Rosenmann, M., and Bozinovic, F. (1989). Energetics and geometry of huddling in small mammals. *Journal of Theoretical Biology*, **141**, 181–9.

Canals, M., Rosenmann, M., and Bozinovic, F. (1997). Geometrical aspects of the energetic effectiveness of huddling in small mammals. *Acta Theriologica*, **42**, 321–8.

Canals, M., Rosenmann, M., Novoa, F. F., and Bozinovic, F. (1998). Modulating factors of the energetic effectiveness of huddling in small animals. *Acta Theriologica*, **43**, 337–48.

Caraco, T. and Wolf, L. L. (1975). Ecological determinants of group size of foraging lions. *American Naturalist*, **109**, 343–52.

Caraco, T., Uetz, G. W., Gillespie, R. G., and Giraldeau, L.-A. (1995). Resource consumption variance within and among individuals: on coloniality in spiders. *Ecology*, **76**, 196–205.

Carbonaro, D. A., Friend, T. H., Dellmeier, G. R., and Nutti, L. C. (1992). Behavioral and physiological responses of dairy goats to isolation. *Physiology and Behavior*, **51**, 297–301.

Carbone, C., Du Toit, J. T., and Gordon, I. J. (1997). Feeding success in African wild dogs: does kleptoparasitism by spotted hyenas influence hunting group size? *Journal of Animal Ecology*, **66**, 318–26.

Carden, S. E. and Hofer, M. A. (1990). Independence of benzodiazepine and opiate action in the suppression of isolation distress in rat pups. *Behavioral Neuroscience*, **104**, 160–6.

Chapman, C. A. and Lefebvre, L. (1990). Manipulating foraging group size – spider monkey food calls in fruiting trees. *Animal Behaviour*, **39**, 891–96.

Cheney, D. L. and Seyfarth, R. M. (1990). *How monkeys see the world*. University of Chicago Press, Chicago.

Chivers, D. P. and Smith, R. J. F. (1994). Fathead minnows, *Pimephales promelas*, acquire predator recognition when alarm substance is associated with the sight of unfamiliar fish. *Animal Behaviour*, **48**, 597–605.

Chivers, D. P., Brown, G. E., and Smith, R. J. F. (1995). Familiarity and shoal cohesion in fathead minnows (*Pimephales promelas*): implications for antipredator behaviour. *Canadian Journal of Zoology*, **73**, 955–60.

Clark, B. R. and Faeth, S. H. (1998). The evolution of egg clustering in butterflies: a test of the egg desiccation hypothesis. *Evolutionary Ecology*, **12**, 543–52.

Clark, C. W. (1987). The lazy, adaptable lions: a Markovian model of group foraging. *Animal Behaviour*, **35**, 361–8.

Clark, C. W. and Mangel, M. (2000). *Dynamic state variable models in ecology: methods and applications*. Oxford University Press, Oxford.

Clotfelter, E. D. and Yasukawa, K. (1999). The effect of aggregated nesting on red-winged blackbird nest success and brood parasitism by brown-headed cowbirds. *Condor*, **101**, 729–36.

Clutton-Brock, T. H. and Harrey, P. H. (1977). Primate ecology and social organisation. *Journal of Zoology London*, **183**, 1–39.

Clutton-Brock, T. H. (1998). Reproductive skew, concessions, and limited control. *Trends in Ecology and Evolution*, **13**, 288–92.

Clutton-Brock, T. H., Price, O. F., and MacColl, A. D. C. (1992). Mate retention, harassment, and the evolution of ungulate leks. *Behavioral Ecology*, **3**, 234–42.

Cockburn, A. (1998). Evolution of helping behaviour in cooperatively breeding birds. *Annual Review of Ecology and Systematics*, **29**, 141–71.

Cook, A. (1981). Huddling and the control of water loss by the slug *Limax pseudoflavus* Evans. *Animal Behaviour*, **29**, 289–98.

Cooper, S. M. (1991). Optimal hunting group size: the need for lions to defend their kills against loss to spotted hyaenas. *African Journal of Ecology*, **29**, 130–6.

Conradt, L. and Roper, T. J. (2001). Activity synchrony and social cohesion: a fission–fusion model. *Proceedings of the Royal Society London Series B*, **267**, 2213–18.

Côté, I. M. and Gross, M. R. (1993). Reduced disease in offspring: a benefit of coloniality in sunfish. *Behavioral Ecology and Sociobiology*, **33**, 269–74.

Côté, I. M. and Poulin, R. (1995). Parasitism and group size in social animals: a meta-analysis. *Behavioral Ecology*, **6**, 159–65.

Couzin, I. D., Krause, J., James, R., Ruxton, G. D., and Franks, N. R. (in press). Collective memory and spatial sorting in animal groups. *Journal of Theoretical Biology*.

Creel, S. (1997). Cooperative hunting and group size: assumptions and currencies. *Animal Behaviour*, **54**, 1319–24.

Creel, S. and Creel, N. M. (1995). Communal hunting and pack size in African wild dogs, *Lycaon pictus*. *Animal Behaviour*, **50**, 1325–39.

Crespi, B. J. (1992). Eusociality in Australian gall thrips. *Nature*, **359**, 724–6.

Crespi, B. J. and Ragsdale, J. E. (2000). A skew model for the evolution of sociality via manipulation: why it is better to be feared than loved. *Proceedings of the Royal Society London Series B*, **267**, 821–8.

Cresswell, W. (1994). Flocking as an effective anti-predation strategy in redshanks, *Tringa totanus*. *Animal Behaviour*, **47**, 433–42.

Cresswell, W. (1998). Variation in the strength of interference competition with resource density in blackbirds, *Turdus merula*. *Oikos*, **81**, 152–60.

Crook, A. C. (1999). Quantitative evidence for assortative schooling in a coral reef fish. *Marine Ecology Progress Series*, **176**, 17–23.

Crook, J. H. (1964). The evolution of social organisation and visual communication in the weaver birds (Ploceinae). *Behaviour Supplement*, **10**, 1–178.

Crook, J. H. and Gartlan, J. S. (1966). Evolution of primate societies. *Nature*, **210**, 1200–3.

Croze, H. (1970). Search image in carrion crows. *Zeitschrift für Tierpsychologie*, **5**, 1–85.

Crozier, R. H. and Pamilo, P. (1996). *Evolution of social insect colonies: sex allocation and kin selection*. Oxford Series in Ecology and Evolution, Oxford University Press, Oxford.

Curio, E. (1978). The adaptive significance of avian mobbing. I. Teleonomic hypotheses and predictions. *Zeitschrift für Tierpsychologie*, **48**, 175–83.

Curio, E., Ernst, U., and Vieth, W. (1978). The adaptive significance of mobbing. II. Cultural transmission of enemy recognition in blackbirds: effectiveness and some constraints. *Zeitschrift für Tierpsychologie*, **48**, 184–202.

Cutts, C. J. and Speakman, J. R. (1994). Energy savings in formation flight of pink-footed geese. *Journal of Experimental Biology*, **189**, 251–61.

Dafni, J. and Diamant, A. (1984). School-oriented mimicry, a new type of mimicry in fishes. *Marine Ecology Progress Series*, **20**, 45–50.

Davies, N. B. (2000). Cuckoos, cowbirds and other cheats. Poyser, London.

Davis, J. M. (1975). Socially induced flight reactions in pigeons. *Animal Behaviour*, **23**, 587–601.

D'Eath, R. B. (1998). Can video images imitate real stimuli in animal behaviour experiments? *Biological Reviews*, **73**, 267–92.

DeBlois, E. M. and Rose, G. A. (1996). Cross-shoal variability in the feeding habits of migrating Atlantic cod (*Gadus morhua*). *Oecologia*, **108**, 192–6.

Dehn, M. M. (1990). Vigilance for predators: detection and dilution effects. *Behavioral Ecology and Sociobiology*, **26**, 337–42.

Deneubourg, J. L., Goss, S., Franks, N. R., and Pasteels, J. M. (1989). The blind leading the blind: modelling chemically mediated army ant raid patterns. *Journal of Insect Behaviour*, **2**, 719–25.

Denno, R. F. and Benrey, B. (1997). Aggregation facilitates larval growth in the neotropical nymphalid butterfly *Chlosyne janais*. *Ecological Entomology*, **22**, 133–41.

Denson, R. D. (1979). Owl predation on a mobbing crow. *Wilson Bulletin*, **91**, 133.

Detrain, C. and Pasteels, J. M. (1992). Caste polyethism and collective defense in the ant *Pheidole pallidula*: the outcome of quantitative differences in recruitment. *Behavioral Ecology and Sociobiology*, **29**, 405–12.

Dexheimer, M. and Southern, W. E. (1974). Breeding success relative to nest location and density in ring-billed gull colonies. *Wilson Bulletin*, **86**, 288–90.

Diamond, J. M. (1981). Mixed-species foraging groups. *Nature*, **292**, 408–9.

Dicke, M. (1986). Volatile spider-mite pheromone and host-plant kairomone, involved in spaced-out gregariousness in the spider mite *Tetranchys urticae*. *Physiological Entomology*, **11**, 251–62.

Dolby, A. S. and Grubb, T. C. Jr. (1998). Benefits to satellite members in mixed species foraging groups: an experimental analysis. *Animal Behaviour*, **56**, 501–9.

Dolby, A. S. and Grubb, T. C. Jr. (1999). Social context affects risk taking by a satellite species in a mixed-species foraging group. *Behavioral Ecology*, **11**, 110–14.

Dolman, P. M. (1995). The intensity of interference varies with resource density: evidence from a field study of snow buntings, *Plectophenax nivalis*. *Oecologia*, **102**, 511–14.

Dominey, W. J. (1981). Anti-predator functions of bluegill sunfish nesting colonies. *Nature*, **290**, 586–7.

Dowdey, T. G. and Brodie, E. D. Jr. (1989). Antipredator strategies of salamanders: individual and geographical variation in responses of *Eurycea bislineata* to snakes. *Animal Behaviour*, **38**, 707–11.

Downs, S. G. and Ratnieks, F. L. W. (1999). Recognition of conspecifics by honeybee guards uses nonheritable cues acquired in the adult stage. *Animal Behaviour*, **58**, 643–8.

Drent, R. and Swierstra, P. (1977). Goose flocks and food-finding: field experiments with barnacle geese in winter. *Wildfowl*, **28**, 15–20.

Driessen, G. and Visser, M. E. (1997). Components of parasitoid interference. *Oikos*, **79**, 179–82.

Duffy, D. C. (1983). The foraging ecology of peruvian seabirds. *Auk*, **100**, 800–10.

Dugatkin, L. A. (1992). Sexual selection and imitation: females copy the mate choice of others. *American Naturalist*, **139**, 1384–9.

Dugatkin, L. A. (1997). *Cooperation among animals: an evolutionary perspective*. Oxford University Press, Oxford.

Dugatkin, L. A. and Alfieri, M. (1991). Guppies and tit-for-tat strategy: preference based on past interaction. *Behavioral Ecology and Sociobiology*, **28**, 243–6.

Dugatkin, L. A. and Godin, J.-G. J. (1992). Reversal of female mate choice by copying in the guppy (*Poecilia reticulata*). *Proceedings of the Royal Society London Series B*, **249**, 179–84.

Dugatkin, L. A. and Höglund, J. (1995). Delayed breeding and the evolution of mate copying in lekking species. *Journal of Theoretical Biology*, **39**, 215–18.

Dugatkin, L. A. and Wilson, D. S. (1992). The prerequisites for strategic behaviour in bluegill sunfish, *Lepomis macrochirus*. *Animal Behaviour*, **44**, 223–30.

Dugatkin, L. A., FitzGerald, G. J., and Lavoir, J. (1994). Juvenile three-spined sticklebacks avoid parasitized conspecifics. *Environmental Biology of Fishes*, **39**, 215–18.

Dukas, R. and Clark, C. W. (1995). Searching for cryptic prey: a dynamic model. *Ecology*, **76**, 1320–6.

Duncan, P. and Vigne, N. (1979). The effect of group size in horses on the rate of attacks by blood-sucking flies. *Animal Behaviour*, **27**, 623–5.

Eggleston, D. B., Lipcius, R. N., Miller, D. L., and Coba-Centina, L. (1990). Shelter scaling regulates survival of juvenile lobster, *Panulirus argus*. *Marine Ecology Progress Series*, **62**, 79–88.

Ehrlich, P. and Ehrlich, A. (1973). Coevolution: heterotypic schooling in Caribbean reef fishes. *American Naturalist*, **107**, 157–60.

Elgar, M. A. (1986). House sparrows establish foraging flocks by giving chirrup calls if the resources are divisible. *Animal Behaviour*, **34**, 169–74.

Elgar, M. A., Burren, P. J., and Posen, M. (1984). Vigilance and perception of flock size in foraging house sparrows (*Passer domesticus* L.). *Behaviour*, **90**, 215–23.

Emlen, S. T. (1991). Evolution of cooperative breeding in birds and mammals. In *Behavioural ecology: an evolutionary approach* (eds. J. R. Krebs and N. B. Davies), pp. 301–37. Blackwell Scientific, Oxford.

Emlen, S. T. (1995). An evolutionary theory of the family. *Proceedings of the National Academy of Sciences of the USA*, **92**, 8092–9.

Endler, J. A. (1991). Interactions between predators and prey. In *Behavioural ecology* (eds. J. R. Krebs and N. B. Davies), pp. 169–201. Blackwell Scientific, Oxford.

Espmark, Y. and Langvatn, R. (1979). Lying down as a means of reducing fly harassment in red deer (*Cervus elaphus*). *Behavioral Ecology and Sociobiology*, **5**, 51–4.

Fanshawe, J. H. and FitzGibbon, C. D. (1993). Factors influencing the hunting success of an African wild dog pack. *Animal Behaviour*, **45**, 479–90.

Fels, D., Rhisiart, A. A., and Vollrath, F. (1995). The selfish crouton. *Behaviour*, **132**, 49–55.

Fields, P. A. (1990). Decreased swimming effort in groups of pacific mackerel (*Scomber japonicus*). *American Society of Zoology*, **30**, 134A.

Fish, F. E. (1991). Energy conservation by formation swimming: metabolic evidence from ducklings. In *Mechanics and physiology of animal swimming* (eds. L. Maddock, Q. Bone, and J. M. V. Rayner). Cambridge University Press, Cambridge.

Fish, F. E. (1995). Kinematics of ducklings swimming in formation: consequences of position. *Journal of Experimental Zoology*, **273**, 1–11.

Fiske, P., Rintamäki, P. T., and Karvonen, E. (1998). Mating success in lekking males: a meta-analysis. *Behavioral Ecology*, **9**, 328–38.

FitzGerald, G. J. and Wootton, R. J. (1993). The behavioural ecology of sticklebacks. In *Behaviour of teleost fishes* (ed. T. J. Pitcher). Chapman & Hall, London.

FitzGibbon, C. D. (1989). A cost to individuals with reduced vigilance in groups of Thomson's gazelles hunted by cheetahs. *Animal Behaviour*, **37**, 508–10.

FitzGibbon, C. D. (1990). Mixed species grouping in Thomson and Grant gazelles—the anti-predator benefits. *Animal Behaviour*, **39**, 1116–26.

Flasskamp, A. (1994). The adaptive significance of avian mobbing. V. An experimental test of the move on hypothesis. *Ethology*, **96**, 322–33.

Flierl, G., Grunbaum, D., Levin, S., and Olson, D. (1999). From individuals to aggregations: the interplay between behavior and physics. *Journal of Theoretical Biology*, **196**, 397–454.

Flynn, R. E. and Giraldeau, L. A. (2001). Producer-scrounger games in a spatially explicit world: Tactic use influences flock geometry of spice finches. *Ethology*, **107**, 249–57.

Foster, J. B. and Dagg, A. I. (1972). Notes of the biology of the giraffe. *East African Wildlife Journal*, **10**, 1–16.

Foster, S. A. (1985). Group foraging by a coral reef fish: a mechanism for gaining access to defended resources. *Animal Behaviour*, **33**, 782–92.

Foster, S. A. (1989). The implications of divergence in spatial nesting patterns in the geminate Caribbean and Pacific sergeant damselfish. *Animal Behaviour*, **37**, 465–76.

Foster, W. A. and Treherne, J. E. (1981). Evidence for the dilution effect in the selfish herd from fish predation on a marine insect. *Nature*, **293**, 466–67.

Frankenberg, E. (1981). The adaptive significance of avian mobbing. IV. Alerting others and perceptual advertisement in blackbirds facing an owl. *Zeitschrift für Tierpsychologie*, **55**, 97–118.

Franks, N. R. (2001). Evolution of mass transit systems in ants: a tale of two societies. In *Insect Movement: Mechanisms and Consequences* (eds. J. Woiwod, C. Thomas and

D. Reynolds), pp. 281–98. *Proceedings of the 20th Symposium of the Royal Entomological Society*, CAB International.

Franks, N. R., Sendova-Franks, A. B., Simmons, J., and Mogie, M. (1999). Convergent evolution, superefficient teams and tempo in Old and New World army ants. *Proceedings of the Royal Society London Series B*, **266**, 1697–701.

Free, C. A., Beddington, J. R., and Lawton, J. H. (1977). On the inadequacy of simple models of mutual interference for parasitism and predation. *Journal of Animal Ecology*, **46**, 543–54.

Fryxell, J. M. (1987). Lek breeding and territorial aggression in white-eared kob. *Ethology*, **75**, 211–20.

Fuiman, L. A. and Magurran, A. E. (1994). Development of predator defences in fishes. *Reviews in Fish Biology and Fisheries*, **4**, 145–83.

Fujita, K., Watanabe, K., Widarto, T. H., and Suryobroto, B. (1997). Discrimination of macaques by macaques: the case of Sulawesi species. *Primates*, **38**, 233–45.

Fullick, T. G. and Greenwood, J. J. D. (1979). Frequency dependent food selection in relation to two models. *American Naturalist*, **113**, 762–5.

Furrer, R. (1975). Häufigkeit und Wirksamkeit des Angriffsverhaltens bei der Wacholderdrossel *Turdus pilaris*. *Ornithologische Beobachtungen*, **72**, 1–8.

Gagliardo, A. and Guilford, T. (1993). Why do aposematic prey live gregariously? *Proceedings of the Royal Society London Series B*, **251**, 69–74.

Gaioni, S. J. and Ross, L. E. (1982). Distress calling induced by reductions in group size in ducklings reared with conspecifics or imprinting stimuli. *Animal Learning and Behavior*, **10**, 521–9.

Gallego, A. and Heath, M. R. (1994). The development of schooling behaviour in Atlantic herring *Clupea harengus*. *Journal of Fish Biology*, **45**, 569–88.

Gamberale, G. and Tullberg, B. S. (1996). Evidence for a more effective signal in aggregated aposematic prey. *Animal Behaviour*, **52**, 597–601.

Gamberale, G. and Tullberg, B. S. (1998). Aposematism and gregariousness: the combined effect of group size and colouration on signal repellence. *Proceedings of the Royal Society London Series B*, **265**, 889–94.

Gatesy, J., Amato, G., Vrba, E., Schaller, G., and DeSalle, R. (1997). A cladistic analysis of mitochondrial ribosomal DNA from the Bovidae. *Molecular Phylogenetics and Evolution*, **3**, 303–19.

Gee, J. H. (1980). Respiratory patterns and anti-predator responses in a central mud minnow, *Umbra limi*, a continuous, facultative air-breathing fish. *Canadian Journal of Zoology*, **58**, 819–27.

Gendron, R. P. (1986). Searching for cryptic prey: evidence for optimal search rates and the formation of search images in quail. *Animal Behaviour*, **34**, 898–912.

Gese, E. M., Rongstad, O. J., and Mytton, W. R. (1988). Relationship between coyote group size and diet in SE Colorado. *Journal of Wildlife Management*, **52**, 647–53.

Gillett, S. D., Hogarth, P. J., and Noble, F. E. J. (1979). The response of predators to varying densities of gregaria locust nymphs. *Animal Behaviour*, **27**, 592–6.

Giraldeau, L.-A. and Beauchamp, G. (1999). Food exploitation: searching for the optimal joining policy. *Trends in Ecology and Evolution*, **14**, 102–6.

Giraldeau, L.-A. and Caraco, T. (2000). *Social Foraging Theory*. Princeton University Press, Princeton.

Giraldeau, L.-A. and Gillis, D. (1985). Optimal group size can be stable: a reply to Sibly. *Animal Behaviour* **33**, 666–7.

Giraldeau, L.-A. and Gillis, D. (1988). Do lions hunt in group sizes that maximise hunters' daily food returns? *Animal Behaviour*, **36**, 611–13.

Godin, J.-G. J. (1986). Risk of predation and foraging behaviour in shoaling banded killifish (*Fundulus diaphanus*). *Canadian Journal of Zoology*, **64**, 1675–8.

Godin, J.-G. J. and Morgan, M. J. (1985). Predator avoidance and school size in a cyprinodontid fish, the banded killifish. *Behavioral Ecology and Sociobiology*, **16**, 105–10.

Godin, J.-G. J., Classon, L. J. and Abrahams, M. V. (1988). Group vigilance and shoal size in a small characin fish. *Behaviour*, **104**, 29–40.

Goodenough, J., McGuire, B., and Wallace, R. A. (1993). *Perspectives in animal behaviour*. Wiley, New York.

Göransson, G., Karlsson, J., Nilsson, S. G., and Ulfstrans, S. (1975). Predation on birds' nests in relation to anti-predator aggression and nest density: an experimental study. *Oikos*, **26**, 117–20.

Gosling, L. M. and Petrie, M. (1990). Lekking in a topi: a consequence of a satellite behaviour by small males at hotspots. *Animal Behaviour*, **40**, 272–87.

Götmark, F., Winkler, D. W., and Andersson, M. (1986). Flock-feeding increases individual success in gulls. *Nature*, **319**, 589–91.

Gould, L. L. and Heppner, F. N. (1974). The vee formation of Canada Geese. *Auk*, **91**, 494–506.

Grant, J. W. A. and Green, L. D. (1995). Mate copying versus preferences for actively courting males by female Japanese medaka (*Orkzias latipes*). *Behavioral Ecology*, **7**, 165–7.

Griffiths, S. W. and Magurran, A. E. (1997a). Familiarity in schooling fish: how long does it take to acquire? *Animal Behaviour*, **53**, 945–9.

Griffiths, S. W. and Magurran, A. E. (1997b). Schooling preferences for familiar fish vary with group size in a wild guppy population. *Proceedings of the Royal Society of London Series B*, **264**, 547–51.

Griffiths, S. W. and Magurran, A. E. (1999). Schooling decisions in guppies (*Poecilia reticulata*) are based on familiarity rather than kin recognition by phenotype matching. *Behavioral Ecology and Sociobiology*, **45**, 437–43.

Groom, M. J. (1992). Sand-coloured nighthawks parasitize the antipredator behaviour of three nesting bird species. *Ecology*, **73**, 785–93.

Gross, M. R. and MacMillan, A. M. (1981). Predation and the evolution of colonial nesting in bluegill sunfish (*Lepomis macrochirus*). *Behavioral Ecology and Sociobiology*, **8**, 163–74.

Gueron, S., Levin, S. A., and Rubenstein, D. I. (1996). The dynamics of herds: From individuals to aggregations. *Journal of Theoretical Biology*, **182**, 85–98.

Gullan, P. J. and Cranston, P. S. (2000). *The insects: an outline of entomology*. Blackwell Science, Oxford.

Gundermann, J. L., Horel, A., and Krafft, B. (1993). Experimental manipulations of social tendencies in the subsocial spider *Coelotes terrestris*. *Insectes Sociaux*, **40**, 219–29.

Hafernik, J. and Saul-Gershenz, L. (2000). Beetle larvae cooperate to mimic bees. *Nature*, **405**, 35.

Hager, M. and Helfman, G. S. (1991). Safety in numbers: shoal size choice by minnows under predatory threat. *Behavioral Ecology and Sociobiology*, **29**, 271–6.

Hainsworth, F. R. (1987). Precision and dynamics of positioning by Canada geese in formation. *Journal of Experimental Biology*, **128**, 445–62.

Hainsworth, F. R. (1988). Induced drag savings from ground effect and formation flight in brown pelicans. *Journal of Experimental Biology*, **135**, 431–44.

Hamilton, I. M. (2000). Recruiters and joiners: using optimal skew theory to predict group size and the division of resources within groups of social animals. *American Naturalist*, **155**, 684–95.

Hamilton, W. D. (1964). The genetical evolution of social behaviour. I, II. *Journal of Theoretical Biology*, **7**, 1–52.

Hamilton, W. D. (1971). Geometry of the Selfish Herd. *Journal of Theoretical Biology*, **31**, 295–311.

Harvey, P. H. (1985). Intrademic group selection and the sex ratio. In *Behavioural ecology: ecological consequences of adaptive behaviour* (eds. R. M. Sibly and R. H. Smith), pp. 59–73. Blackwell Scientific Publications, Oxford.

Harvey, P. H. and Pagel, M. D. (1991). *The comparative method in evolutionary biology*. Oxford University Press, Oxford.

Hatchwell, B. J. and Komdeur, J. (2000). Ecological constraints, life history traits and the evolution of cooperative breeding. *Animal Behaviour*, **59**, 1079–86.

Hauser, M. D. (1992). Costs of deception: cheaters are punished in rhesus monkeys (*Macaca mulatta*). *Proceedings of the National Academy of Sciences of the USA*, **89**, 12127–39.

Hauser, M. D. (2001). Searching for food in the wild: a nonhuman primate's expectations about invisible displacement. *Developmental Science*, **4**, 84–93.

Hayes, J. P., Speakman, J. R., and Racey, P. A. (1992). The contributions of local heating and reducing exposed surface area to the energetic benefits of huddling by Short-tailed Field Voles. *Physiological Zoology*, **65**, 742–62.

Hector, D. P. (1986). Cooperative hunting and its relationship to foraging success and prey size in an avian predator. *Ethology*, **73**, 247–57.

Heinrich, B. and Marzluff, J. M. (1991). Do common ravens yell because they want to attract others? *Behavioral Ecology and Sociobiology*, **28**, 13–21.

Heinrich, B. and Marzluff, J. M. (1995). Why ravens share. *American Scientist*, **83**, 342–9.

Heinrich, B., Marzluff, J. M., and Marzluff, C. S. (1993). Common ravens are attracted by appeasement calls of food discoverers when attacked. *Auk*, **110**, 247–54.

Helle, T. and Aspi, J. (1983). Does herd formation reduce insect harassment among reindeer? A field experiment with animal traps. *Acta Zoologica Fennica*, **175**, 129–31.

Hemelrijk, C. K. (2000). Towards the integration of social dominance and spatial structure. *Animal Behaviour*, **59**, 1035–48.

Hennesy, M. B., Becker, L. A., and O'Neil, D. R. (1991). Peripherally administered CRH suppresses the vocalizations of isolated guinea pig pups. *Physiology and Behavior*, **50**, 17–22.

Heppner, F. H. (1997). Three-dimensional structure and dynamics of bird flocks. In *Animal groups in three dimensions* (eds. J. K. Parrish and W. M. Hamner), pp. 68–87. Cambridge University Press, Cambridge.

Heppner, F. H., Convissar, J. L., Moonan, D. E. Jr., and Anderson, J. G. T. (1985). Visual angle and formation flight in Canada Geese (*Branta canadesis*). *Auk*, **102**, 195–8.

Hepper, P. G. and Cleland, J. (1998). Developmental aspects of kin recognition. *Genetica*, **104**, 199–205.

Herskin, J. and Steffensen, J. F. (1998). Energy savings in sea bass swimming in a school: measurements of tail beat frequency and oxygen consumption at different swimming speeds. *Journal of Fish Biology*, **53**, 366–76.

Herzog, H. A. and Schwartz, J. M. (1990). Geographical variation in anti-predator behavior of neonate garter snakes, *Thamnophis sirtalis*. *Animal Behaviour*, **40**, 597–601.

Higashi, M. and Yamamura, N. (1993). What determines the animal group size? Insider–outsider conflict and its resolution. *American Naturalist*, **142**, 553–63.

Higdon, J. J. L. and Corrsin, S. (1978). Induced drag in a bird flock. *American Naturalist*, **112**, 727–44.

Hilborn, R. (1991). Modelling the stability of fish schools: exchange of individual fish between schools of skipjack tuna *Katsuwonus pelamis*. *Canadian Journal of Fisheries and Aquatic Sciences*, **48**, 1081–91.

Hildrew, A. G. and Townsend, C. R. (1980). Aggregation, interference and foraging by larvae of *Plectrocnemia conspersa*. *Animal Behaviour*, **28**, 553–60.

Hirth, D. H. and McCullough, D. R. (1977). Evolution of alarm signals in ungulates with special reference to white-tailed deer. *American Naturalist*, **111**, 31–42.

Hoare, D. J., Krause, J., Ruxton, G. D., and Godin, J.-G. J. (2000). The social organisation of free-ranging fish shoals. *Oikos*, **89**, 546–54.

Höglund, J. and Alatalo, R. V. (1995). *Leks*. Princeton University Press, Princeton.

Höglund, J., Alatalo, R. V., Lundberg, A., Rintamäki, P. T., and Lindell, J. (1999). Microsatellite markers reveal the potential for kin selection on black grouse leks. *Proceedings of the Royal Society London Series B*, **266**, 813–16.

Hoi, H., Darolova, A., Konig, C., and Kristofik, J. (1998). The relation between colony size, breeding density and ectoparasite load in adult bee-eaters. *Ecoscience*, **5**, 156–63.

Höjesjö, J., Johnsson, J. L., Petersson, E., and Järvi, T. (1998). The importance of being familiar: individual recognition and social behavior in sea trout (*Salmo trutta*). *Behavioral Ecology*, **9**, 445–51.

Hölldobler, B. and Wilson, O. E. (1990). *The ants*. Harvard University Press, Cambridge.

Holmgren, N. (1995). The ideal free distribution of unequal competitors: predictions from a behaviour-based model. *Journal of Animal Ecology*, **64**, 197–212.

Hoogland, J. L. (1979). Aggression, ectoparasitism, and other possible costs of prairie dog coloniality. *Behaviour*, **69**, 1–35.

Hoogland, J. L. and Sherman, P. W. (1976). Advantages and disadvantages of bank swallow (*Riparia riparia*) coloniality. *Ecological Monographs*, **46**, 33–58.

Horn, H. S. (1968). The adaptive significance of colonial nesting in the Brewer's blackbird (*Euphagus cyanocephalus*). *Ecology*, **49**, 682–94.

Houston, A. I., Clark, C. W., McNamara, J. M., and Mangel, M. (1988). Dynamic models in behavioural and evolutionary ecology. *Nature*, **332**, 29–34.

Huang, Z. Y., Plettner, E., and Robinson, G. E. (1998). Effects of social environment and worker mandibular glands on endocrine-mediated behavioral development in honey bees. *Journal of Comparative Physiology A*, **183**, 143–52.

Hummel, D. (1983). Aerodynamic aspects of formation flight in birds. *Journal of Theoretical Biology*, **104**, 321–47.

Hummel, D. (1995). Formation flight as an energy-saving mechanism. *Israel Journal of Zoology*, **41**, 261–78.

Hunter, J. R. (1966). Procedure for analysis of schooling behaviour. *Journal of the Fisheries Research Board of Canada*, **23**, 547–62.

Huntingford, F. A., Wright, P. J., and Tierney, J. F. (1994). Adaptive variation in antipredator behaviour in threespine stickleback. In *The evolutionary biology of the threespine stickleback* (eds. M. A. Bell and S. A. Foster), pp. 277–96. Oxford University Press, Oxford.

Huth, A. and Wissel, C. (1992). The simulation of the movement of fish schools. *Journal of Theoretical Biology*, **156**, 365–85.

Ikawa, T. and Okabe, H. (1997). Three-dimensional measurements of swarming mosquitoes: a probabilistic model, measuring system, and example results. In *Animal groups in three dimensions* (eds. J. K. Parrish and W. M. Hamner), pp. 90–104. Cambridge University Press, Cambridge.

Inglis, I. R. and Isaacson, A. J. (1978). The responses of dark-bellied brent geese to models of geese in various postures. *Animal Behaviour*, **26**, 953–8.

Inman, A. I. and Krebs, J. R. (1987). Predation and group living. *Trends in Ecology and Evolution*, **2**, 31–2.

Jakobsen, P. J. and Johnsen, G. H. (1988). Size-specific protection against predation by fish in swarming waterfleas, *Bosmina longispina*. *Animal Behaviour*, **36**, 986–90.

Janetos, A. C. (1980). Strategies of female choice: a theoretical analysis. *Behavioral Ecology and Sociobiology*, **7**, 107–12.

Janik, V. M. (2000). Food-related bray calls in wild bottlenose dolphins (*Tursiops truncatus*). *Proceedings of the Royal Society London Series B*, **267**, 923–7.

Janson, C. H. (1985). Aggressive competition and individual food consumption in the brown capuchin monkey (*Cebus apella*). *Behavioral Ecology and Sociobiology*, **18**, 125–38.

Janson, C. H. (1990). Ecological consequences of individual spatial choice in foraging groups of brown capuchin monkeys, *Cebus apella*. *Animal Behaviour*, **40**, 922–34.

Jarman, P. J. (1974). The social organization of antelope in relation to their ecology. *Behaviour*, **48**, 215–67.

Johnstone, R. A. and Earn, D. J. D. (1999). Imperfect female choice and male mating skew on leks of different sizes. *Behavioral Ecology and Sociobiology*, **45**, 277–81.

Judd, T. M. and Sherman, P. W. (1996). Naked mole-rates recruit colony mates to food sources. *Animal Behaviour*, **52**, 957–69.

Keenleyside, M. H. A. (1955). Some aspects of the schooling behaviour of fish. *Behaviour*, **8**, 183–248.

Keller, L. and Reeve, H. K. (1994). Partitioning of reproduction in animal societies. *Trends in Ecology and Evolution*, **9**, 98–102.

Kennedy, J. S. and Crawley, L. (1967). Spaced-out gregariousness in sycamore aphids *Depanosiphum platanoides* (Schrank) (Hemiptera, Callaphidae). *Journal of Animal Ecology*, **36**, 147–63.

Kent, D. S. and Simpson, J. A. (1992). Eusociality in the beetle *Austroplatypus incompertus* (Coleoptera: Curculionidae). *Naturwissenschaften*, **79**, 86–7.

Kenwood, R. E. (1978). Hawks and doves: attack success and selection in goshawk flights at wood-pigeons. *Journal of Animal Ecology*, **47**, 449–60.

Keys, G. C. and Dugatkin, L. A. (1990). Flock size and position effects on vigilance, aggression, and prey capture in the European starling. *Condor*, **92**, 151–9.

Kirkpatrick, M. and Dugatkin, L. A. (1994). Sexual selection and the evolutionary effects of matecopying. *Behavioral Ecology and Sociobiology*, **34**, 443–9.

Klok, C. J. and Chown, S. L. (1999). Assessing the benefits of aggregation: thermal biology and water relations of anomalous Emperor Moth caterpillars. *Functional Ecology*, **13**, 417–19.

Klump, G. M. and Shalter, M. D. (1984). Acoustic behaviour of birds and mammals in the predator context. *Zeitschrift für Tierpsychologie*, **66**, 189–226.

Kokko, H. (1997). The lekking game: can female choice explain aggregated male displays? *Journal of Theoretical Biology*, **187**, 57–64.

Kokko, H. and Johnstone, R. A. (1999). Social queuing in animal societies: a dynamic model of reproductive skew. *Proceedings of the Royal Society London Series B*, **266**, 571–8.

Kokko, H. and Lindström, J. (1996). Kin selection and the evolution of leks: whose success do young males maximize? *Proceedings of the Royal Society London Series B*, **263**, 919–23.

Kokko, H., Sutherland, W. J., Lindström, J., Reynolds, J. D., and Mackenzie, A. (1998). Individual mating success, lek stability, and the neglected limitations of statistical power. *Animal Behaviour*, **56**, 755–62.

Kokko, H., Rintamäki, P. T., Alatalo, R. V., Höglund, J., Karvonen, E., and Lundberg, A. (1999a). Female choice selects for lifetime lekking performance in black grouse males. *Proceedings of the Royal Society London Series B*, **266**, 2109–15.

Kokko, H., Mackenzie, A., Reynolds, J. D., Lindström, J., and Sutherland, W. J. (1999b). Measures of inequality are not equal. *American Naturalist*, **154**, 358–82.

Kokko, H., Johnstone, R. A., and Clutton-Brock, T. H. (2000). The evolution of cooperative breeding through group augmentation. *Proceedings of the Royal Society London Series B*, **268**, 187–96.

Komdeur, J. (1992). Importance of habitat saturation and territory quality for the evolution of cooperative breeding in the Seychelles Warbler. *Nature*, **358**, 493–5.

Krafft, B., Horel, A., and Julita, J. (1986). Influence of food supply on the duration of the gregarious phase of a maternal-social spider, *Coelotes terrestris* (*Araneae, Agelenidae*). *Journal of Arachnology*, **14**, 219–26.

Krakauer, D. C. (1995). Groups confuse predators by exploiting perceptual bottlenecks: a connectionist model of the confusion effect. *Behavioral Ecology and Sociobiology*, **36**, 421–9.

Kramer, D. L. (1995). Are colonies superoptimal groups? *Animal Behaviour*, **33**, 1031–2.

Kramer, B. and Kuhn, B. (1994). Species recognition by the sequence of discharge intervals in weakly electric fishes of the genus *Campylormyrus* (Mormyridae, Teleostei). *Animal Behaviour*, **48**, 435–45.

Krause, J. (1993a). Transmission of fright reaction between different species of fish. *Behaviour*, **127**, 37–48.

Krause, J. (1993b). The relationship between foraging and shoal position in a mixed shoal of roach (*Rutilus rutilus*) and chub (*Leuciscus leuciscus*): a field study. *Oecologia*, **93**, 356–9.

Krause, J. (1993c). The effect of 'Schreckstoff' on the shoaling behaviour of the minnow—a test of Hamilton's selfish herd theory. *Animal Behaviour*, **45**, 1019–24.

Krause, J. (1994a). The influence of food competition and predation risk on size-assortative shoaling in juvenile chub (*Leuciscus cephalus*). *Ethology*, **96**, 105–16.

Krause, J. (1994b). Differential fitness returns in relation to spatial positions in groups. *Biological Reviews*, **69**, 187–206.

Krause, J. and Godin, J.-G. J. (1994a). The influence of parasitism on the shoaling behaviour of the killifish (*Fundulus diaphanus*). *Canadian Journal of Zoology*, **72**, 1775–9.

Krause, J. and Godin, J.-G. J. (1994b). Shoal choice in banded killifish (*Fundulus diaphanus*, Teleostei, Cyprinodontidae): The effects of predation risk, fish size, species composition and size of shoals. *Ethology*, **98**, 128–36.

Krause, J. and Godin, J.-G. J. (1995). Predator preferences for attacking particular group sizes: consequences for predator hunting success and prey predation risk. *Animal Behaviour*, **50**, 465–73.

Krause, J. and Godin, J.-G. J. (1996a). Influence of prey foraging posture on predator detectability and predation risk: predators take advantage of unwary prey. *Behavioral Ecology*, **7**, 264–71.

Krause, J. and Godin, J.-G. J. (1996b). Influence of parasitism on shoal choice in the banded killifish (*Fundulus diaphanus*, Teleostei, Cyprinodontidae). *Ethology*, **102**, 40–9.

Krause, J. and Tegeder, R. W. (1994). The mechanism of aggregation behaviour in fish shoals: individuals minimise approach time to neighbours. *Animal Behaviour*, **48**, 353–9.

Krause, J., Bumann, D., and Todt, D. (1992). Relationship between the position preference and nutritional state of individuals in schools of juvenile roach (*Rutilus rutilus*). *Behavioral Ecology and Sociobiology*, **30**, 177–80.

Krause, J., Godin, J.-G. J., and Brown, D. (1996a). Phenotypic variability within and between fish shoals. *Ecology*, **77**, 1586–91.

Krause, J., Godin, J.-G. J., and Brown, D. (1996b). Size-assortativeness in multi-species fish shoals. *Journal of Fish Biology*, **49**, 221–5.

Krause, J., Rubenstein, D. I., and Brown, D. (1997). Shoal choice behaviour in fish: the relationship between assessment time and assessment quality. *Behaviour*, **134**, 1051–62.

Krause, J., Ruxton, G. D., and Rubenstein, D. I. (1998a). Is there an influence of group size on predator hunting success? *Journal of Fish Biology*, **52**, 494–501.

Krause, J., Reeves, P., and Hoare, D. (1998b). Positioning behaviour in roach shoals: the role of body length and nutritional state. *Behaviour*, **135**, 1031–9.

Krause, J., Godin, J.-G. J., and Brown, D. (1998c). Body size variation among multi-species fish shoals: the effects of shoal size and number of species. *Oecologia*, **114**, 67–72.

Krause, J., Hoare, D. J., Croft, D., Lawrence, J., Ward, A. W., James, R., *et al.* (2000a). Fish shoal composition: mechanisms and constraints. *Proceedings of the Royal Society London Series B*, **267**, 2011–17.

Krause, J., Butlin, R. K., Peuhkuri, N., and Pritchard, V. L. (2000b). The social organization of fish shoals: a test of the predictive power of laboratory experiments for the field. *Biological Reviews*, **75**, 477–501.

Krebs, J. R. (1971). Territory and breeding density in the great tit, *Parus major. Ecology*, **52**, 2–22.

Krebs, J. R. (1974). Colonial nesting and social feeding as strategies for exploiting food resources in the great blue heron (*Ardea herodias*). *Behaviour*, **51**, 99–134.

Krebs, J. R. and Davies, N. B. (1993). *An introduction to behavioural ecology*, 3rd edn. Blackwell Scientific Publications, Oxford.

Krebs, J. R. and Kacelnik, A. (1991). Decision-making. In *Behavioural ecology: an evolutionary approach* (eds. J. R. Krebs and N. B. Davies), pp. 105–36. Blackwell Scientific Publications, Oxford.

Krebs, J. R., MacRoberts, M. H., and Cullen, J. M. (1972). Flocking and feeding in the great tit *Parus major*, an experimental study. *Ibis*, **114**, 507–30.

Kristiansen, J. N., Fox, A. D., Boyd, H., and Stround, D. A. (2000). Greenland white-fronted geese *Anser aibifrons flavirostris* benefit from feeding in mixed-species flocks. *Ibis*, **142**, 1142–4.

Kruuk, H. (1964). Predators and anti-predator behaviour of the black-headed gull (*Larus ridibundus* L.). *Behaviour Supplements*, **11**, 1–129.

Kshatriya, M. and Blake, R. W. (1992). Theoretical model of optimal flock size of birds flying in formation. *Journal of Theoretical Biology*, **157**, 135–74.

Kunin, W. E. (1999). Patterns of herbivore incidence on experimental arrays and field populations of ragwort, *Senecio jacobaea. Oikos*, **84**, 515–25.

Kunzler, R. and Bakker, T. C. M. (1998). Computer animations as a tool in the study of mating preferences. *Behaviour*, **135**, 1137–59.

Kurta, A. and Fujita, M. S. (1988). Design and interpretation of laboratory thermoregulation studies. In *Ecological and behavioral methods for the study of bats* (ed. T. H. Kunz), pp. 333–52. Smithsonian Institution Press, Washington.

Kyle, C. R. (1979). Reduction in wind resistance and power output of racing cyclists and runers travelling in groups. *Ergonomics*, **22**, 387–97.

Lachlan, R. F., Crooks, L., and Laland, K. N. (1998). Who follows whom? Shoaling preferences and social learning of foraging information in guppies. *Animal Behaviour*, **56**, 181–90.

Lafferty, K. D. and Morris, A. K. (1996). Altered behavior of parasitised killifish increases susceptibility to predation by bird final hosts. *Ecology*, **77**, 1390–7.

Lafleur, D. L., Lozano, G. A., and Sclafani, M. (1997). Female mate-choice copying in guppies, *Poecillia reticulata*: a re-evaluation. *Animal Behaviour*, **54**, 579–86.

Laland, K. N. (1994a). On the evolutionary consequences of sexual imprinting. *Evolution*, **48**, 477–89.

Laland, K. N. (1994b). Sexual selection with a culturally-transmitted mating preference. *Theoretical Population Biology*, **45**, 1–15.

Laland, K. N. and Reader, S. M. (1999). Foraging innovation is inversely related to competitive ability in male but not in female guppies. *Behavioral Ecology*, **10**, 270–4.

Laland, K. N. and Williams, K. (1997). Shoaling generates social learning of foraging information in guppies. *Animal Behaviour*, **53**, 1161–9.

Laland, K. N. and Williams, K. (1998). Social transmission of maladaptive information in the guppy. *Behavioral Ecology*, **9**, 493–9.

Laland, K. N., Odling-Smee, J., and Feldman, M. W. (2000). Niche construction earns its keep. *Behaviour and Brain Sciences*, **23**, 163–74.

Lancaster, J. (1971). Play-mothering: the relations between juvenile females and young infants among free-ranging vervet monkeys (*Cercopithecus aethiops*). *Folia Primatologica*, **15**, 161–82.

Landeau, L. and Terborgh, J. (1986). Oddity and the confusion effect in predation. *Animal Behaviour*, **34**, 1372–80.

Larsen, T. (2000). Influence of rodent density on nesting associations involving bar-tailed godwit *Limosa tapponia*. *Ibis*, **142**, 476–481.

Launay, F., Mills, A. D., Faure, J.-M., and Williams, J. B. (1993). Effects of CRF on isolated Japanese quails selected for fearfulness and for sociality. *Physiology and Behavior*, **54**, 111–18.

Lazarus, J. (1979). The early warning function of flocking in birds: an experimental study with captive quelea. *Animal Behaviour*, **27**, 855–65.

Le Conte, Y., Mohammedi, A., and Robinson, G. E. (2001). Primer effects of a brood pheromone on honeybee behavioural development. *Proceedings of the Royal Society London Series B*, **268**, 163–8.

Lee, P. C. (1994). Social structure and evolution. In *Behaviour and evolution* (eds. P. J. B. Slater and T. R. Halliday), pp. 266–303. Cambridge University Press, Cambridge.

Le Masurier, A. D. (1994). Costs and benefits of egg clustering in *Pieris brassicae*. *Journal of Animal Ecology*, **63**, 677–85.

Lemly, A. D. and Esch, G. W. (1984). Effects of the trematode *Ulifer amboplitis* on juvenile bluegill sunfish, *Lepomis macrochirus*: ecological implications. *Journal of Parasitology*, **70**, 475–92.

Levin, S. A. (1997). Conceptual and methodological issues in the modelling of biological aggregations. In *Animal groups in three dimensions* (eds. J. K. Parrish and W. M. Hamner), pp. 247–56. Cambridge University Press, Cambridge.

Lima, S. L. (1994a). On the personal benefits of anti-predatory vigilance. *Animal Behaviour*, **48**, 734–6.

Lima, S. L. (1994b). Collective detection of predatory attacks by birds in the absence of alarm signals. *Journal of Avian Biology*, **25**, 319–26.

Lima, S. L. (1995a). Back to the basics of anti-predatory vigilance: the group size effect. *Animal Behaviour*, **49**, 11–20.

Lima, S. L. (1995b). Collective detection of predatory attack by social foragers: fraught with ambiguity. *Animal Behaviour*, **50**, 1097–108.

Lima, S. L. and Bednekoff, P. A. (1999). Back to the basics of anti-predatory vigilance: can nonvigilant animals detect attack? *Animal Behaviour*, **58**, 537–43.

Lindström, Å. (1989). Finch flock size and the risk of hawk predation at a migratory stopover site. *Auk*, **106**, 225–32.

Lindström, L. (1999). Experimental approaches to studying the initial evolution of conspicuous aposematic signalling. *Evolutionary Ecology*, **13**, 605–18.

Lindström, K. and Ranta, E. (1993). Foraging group structure among individuals differing in competitive ability. *Annales Zoologici Fennici*, **30**, 225–32.

Linton, M. C., Crowley, P. H., Williams, J. T., Dillon, P. M., Strohmeier, K. L., and Wood, C. (1991). Pit relocation by antlion larvae: a simple model and laboratory test. *Evolutionary Ecology*, **5**, 93–104.

Lissaman, P. B. S. and Shollenberger, C. A. (1970). Formation flight in birds. *Science*, **168**, 1003–5.

Lloyd, M. and Dybas, H. S. (1966). The periodical cicada problem. II. Evolution. *Evolution*, **20**, 466–505.

Loehle, C. (1995). Social barriers to pathogen transmission in wild animal populations. *Ecology*, **76**, 326–35.

Lott, D. F. and Minta, S. C. (1983). Random individual association and social group instability in American bison (*Bison bison*). *Zeitschrift für Tierpsychologie*, **61**, 153–72.

Loughry, W. J. (1988). Population differences in how black-tailed prairie dogs deal with snakes. *Behavioral Ecology and Sociobiology*, **22**, 627–30.

Lynch, L. D. (1998). Indirect mutual interference and the CV2 > 1 rule. *Oikos*, **82**, 318–26.

Mackenzie, A., Reynolds, J. D., Brown, V. J., and Sutherland, W. J. (1995a). Variation in male mating success on leks. *American Naturalist*, **145**, 633–52.

Magurran, A. E. (1986). Predator inspection in minnow shoals: differences between populations and individuals. *Behavioral Ecology and Sociobiology*, **19**, 267–73.

Magurran, A. E. (1990). The inheritance and development of minnow anti-predator behaviour. *Animal Behaviour*, **39**, 834–42.

Magurran, A. E. and Higham, A. (1988). Information transfer across fish shoals under predation threat. *Ethology*, **78**, 153–8.

Magurran, A. E. and Pitcher, T. J. (1987). Provenance, shoal size and the sociobiology of predator-evasion behaviour in minnow shoals. *Proceedings of the Royal Society London Series B*, **229**, 439–65.

Magurran, A. E. and Seghers, B. H. (1990). Population differences in the schooling behaviour of newborn guppies, *Poecilia reticulata*. *Ethology*, **84**, 334–42.

Magurran, A. E. and Seghers, B. H. (1991). Variation in schooling and aggression amongst guppy (*Poecilia reticulata*) populations in Trinidad. *Behaviour*, **118**, 214–34.

Magurran, A. E. and Seghers, B. H. (1994). Predator inspection behaviour covaries with schooling tendency amongst wild guppy, *Poecilia reticulata*, populations in Trinidad. *Behaviour*, **128**, 121–34.

Magurran, A. E., Seghers, B. H., Carvalho, G. R., and Shaw, P. W. (1992). Behavioural consequences of an artificial introduction of guppies (*Poecilia reticulata*) in N. Trinidad:

evidence for the evolution of antipredator behaviour in the wild. *Proceedings of the Royal Society London Series B*, **248**, 117–22.

Magurran, A. E., Seghers, B. H., Shaw, P. W., and Carvalho, G. R. (1994). Schooling preferences for familiar fish in the guppy, *Poecilia reticulata*. *Journal of Fish Biology*, **45**, 401–6.

Magurran, A. E., Seghers, B. H., Shaw, P. W., and Carvalho, G. R. (1995). The behavioral diversity and evolution of guppy, *Poecilia reticulata*, populations in Trinidad. *Advances in the Study of Behavior*, **24**, 155–202.

Major, P. F. (1978). Predator-prey interactions in two schooling fishes, *Caranx ignobilis* and *Stolephorus purpureus*. *Animal Behaviour*, **26**, 760–77.

Mangel, M. and Clark, C. W. (1988). *Dynamic modelling in behavioural ecology*. Princeton University Press, Princeton.

Marler, P. and Peters, S. (1977). Selective vocal learning in a sparrow. *Science*, **198**, 519–21.

Marler, P., Dufty, A., and Pickert, R. (1986). Vocal communication in the domestic chicken. 1. Does the sender communicate information about the quality of a food referent to a receiver? *Animal Behaviour*, **34**, 188–93.

Masuda, R. and Tsukamoto, K. (1998). The ontogeny of schooling behaviour in the striped jack. *Journal of Fish Biology*, **52**, 483–93.

Mateo, J. M. and Johnston, R. E. (2000). Retention of social recognition after hibernation in Belding's ground squirrels. *Animal Behaviour*, **59**, 491–9.

Mather, J. A. and O'Dor, R. K. (1984). Spatial organization of schools of the squid *Illex illecebrosus*. *Marine Behavioural Physiology*, **10**, 259–71.

Mathis, A., Chivers, D. P., and Smith, R. J. F. (1996). Cultural transmission of predator recognition in fishes: intraspecific and interspecific learning. *Animal Behaviour*, **51**, 185–201.

Maynard Smith, J. (1976). Evolution and the theory of games. *American Scientific*, **64**, 41–5.

Maynard Smith, J. (1998). The origin of altruism. *Nature*, **393**, 639–40.

McCaffery, A. R., Simpson, S. J., Islam, M. S., and Roessingh, P. (1998). A gregarizing factor present in the egg pod foam of the desert locust *Schistocera gregaria*. *Journal of Experimental Biology*, **201**, 347–63.

McCann, L. I., Koehn, D. J., and Kline, N. J. (1971). The effects of body-size and body markings on nonpolarized schooling behavior of zebra fish (*Brachydanio rerio*). *Journal of Psychology*, **79**, 71–5.

McNamara, J. M. and Houston, A. I. (1992). Evolutionarily stable levels of vigilance as a function of group size. *Animal Behaviour*, **43**, 641–58.

McRoberts, S. P. and Bradner, J. (1998). The influence of body coloration on shoaling preferences in fish. *Animal Behaviour*, **56**, 611–15.

Metcalfe, N. B. (1984). The effects of mixed-species flocking on the vigilance of shorebirds: who do they trust? *Animal Behaviour*, **32**, 986–93.

Metcalfe, N. B. (1989). Flocking preferences in relation to vigilance benefits and aggression costs in mixed-species shorebird flocks. *Oikos*, **56**, 91–8.

Metcalfe, N. B. and Thomson, B. C. (1995). Fish recognize and prefer to shoal with poor competitors. *Proceedings of the Royal Society of London Series B*, **259**, 207–10.

Milinski, M. (1977a). Experiments on the selection by predators against spatial oddity of their prey. *Zeitschrift für Tierpsychologie*, **43**, 311–25.

Milinski, M. (1977b). Do all members of a swarm suffer the same predation? *Zeitschrift für Tierpsychologie*, **45**, 373–88.

Milinski, M. (1987). Tit-for-tat in sticklebacks and the evolution of cooperation. *Nature*, **325**, 433–7.

Milinski, M. (1990). Information overload and food selection. In *Behavioural mechanisms of food selection* (ed. R. N. Hughes). Springer–Verlag, Berlin.

Milinski, M. and Heller, R. (1978). Influence of a predator on the optimal foraging behaviour of sticklebacks (*Gasterosteus aculeatus* L.). *Nature*, **275**, 642–4.

Milinski, M., Külling, D., and Kettler, R. (1990). Tit for tat: sticklebacks 'trusting' a cooperating partner. *Behavioral Ecology*, **1**, 7–12.

Mills, A. D. and Faure, J. M. (1990). The treadmill test for the measurement of social motivation in Phsianidae chicks. *Medical Science Research*, **18**, 179–80.

Mock, D. W., Lamey, T. C., and Thompson, D. B. A. (1988). Falsifiability and the information centre hypothesis. *Ornis Scandinavia*, **19**, 231–48.

Moller, A. P. (1987). Advantages and disadvantages of coloniality in swallows, *Hirundo rustica*. *Animal Behaviour*, **35**, 819–32.

Moller, A. P. and Birkhead, A. P. (1993). Cuckoldry and sociality: a comparative study of birds. *American Naturalist*, **142**, 118–40.

Moore, J., Simberloff, D., and Freehling, M. (1988). Relationships between bobwhite quail social-group size and intestinal helminth parasitism. *American Naturalist*, **131**, 22–32.

Mooring, M. S. and Hart, B. J. (1992). Animal grouping for protection from parasites: selfish herd and encounter-dilution effects. *Behaviour*, **123**, 173–93.

Morgan, M. J. and Colgan, P. W. (1987). The effects of predator presence and shoal size on foraging in bluntnose minnows, *Pimephales notatus*. *Environmental Biology of Fishes*, **20**, 105–11.

Morton, T. L., Haefner, J. W., Nugala, V., Decino, R. D., and Mendes, L. (1994). The selfish herd revisited: do simple movement rules reduce relative predation risk? *Journal of Theoretical Biology*, **167**, 73–79.

Munn, C. A. (1986). Birds that 'cry wolf'. *Nature*, **319**, 143–5.

Murdoch, W. W. and Stewart-Oaten, A. (1975). Predation and population stability. *Advances in Ecological Research*, **9**, 2–131.

Murton, R. K., Isaacson, A. J., and Westwood, N. J. (1966). The relationship between wood-pigeons and their clover food supply and the mechanism of population control. *Journal of Applied Ecology*, **3**, 55–94.

Murton, R. K., Isaacson, A. J., and Westwood, N. J. (1971). The significance of gregarious feeding behaviour and adrenal stress in a population of wood pigeons *Columba palumbus*. *Journal of Zoology*, **165**, 53–84.

Murton, R. K., Westwood, N. J., and Isaacson, A. J. (1974). A study of wood pigeon shooting: the exploitation of a natural population. *Journal of Applied Ecology*, **11**, 61–81.

Neill, S. R. St. J. and Cullen, J. M. (1974). Experiments on whether schooling of prey affects hunting behaviour of cephalopods and fish predators. *Journal of Zoology*, **172**, 549–69.

Newson, R. M., Mella, P. N. P., and Franklin, T. E. (1973). Observations on the numbers of the tick *Rhipicephalus appendiculatus* on the ears of zebu cattle in relation to hierarchical status in the herd. *Tropical Animal Health and Production*, **5**, 281–3.

Norberg, U. M. (1989). *Vertebrate flight*. Springer–Verlag, Berlin, Heidelberg, New York.

Nordell, S. E. and Valone, T. J. (1998). Mate choice copying as public information. *Ecology Letters*, **1**, 74–6.

O'Connell, C. P. (1972). The interrelationship of biting and filter feeding activity of the northern anchovy (*Engraulis mordax*). *Journal of the Fisheries Research Board of Canada*, **29**, 285–93.

Ohguchi, O. (1978). Experiments on the selection against colour oddity of water fleas by three-spined sticklebacks. *Zeitschrift für Tierpsychologie*, **47**, 254–67.

Ohguchi, O. (1981). Prey density and selection against oddity by three-spined sticklebacks. *Zeitschrift für Tierpsychologie, Supplements*, **23**, 1–79.

Okabe, A., Boots, B. N., and Sugihara, K. (1992). *Spatial tesselations: concepts and applications of Voronoi diagrams*. Wiley, Chichester.

Okasha, S. (2001). Why won't the group selection controversy go away? *British Journal for the Philosophy of Science*, **52**, 25–50.

Okamura, B. (1986). Group living and the effects of spatial position in aggregations of *Mytilus edulis*. *Oecologia*, **69**, 341–7.

Okubo, A. (1986). Dynamic aspects of animal grouping: swarms, schools, flocks, and herds. *Advances in Biophysics*, **22**, 1–94.

Olsén, K. H. (1989). Sibling recognition in juvenile Arctic charr (*Salvelinus alpinus* (L.)). *Journal of Fish Biology*, **34**, 571–81.

Olsén, K. H., Grahn, M., Lohm, J., and Langefors, A. (1998). MHC and kin discrimination in juvenile arctic charr, *Salvelinus alpinus* (L.). *Animal Behaviour*, **56**, 319–27.

Packer, C. (1977). Reciprocal altruism in *Papio anubis*. *Nature*, **265**, 441–3.

Packer, C. and Abrams, P. (1990). Should co-operative groups be more vigilant than selfish ones? *Journal of Theoretical Biology*, **142**, 341–57.

Packer, C. and Caro, T. M. (1997). Foraging costs in social carnivores. *Animal Behavior*, **54**, 1317–18.

Packer, C., Scheel, D., and Pusey, A. E. (1990). Why lions form groups: food is not enough. *American Naturalist*, **136**, 1–19.

Page, G. and Whitacre, D. F. (1975). Raptor predation on wintering shorebirds. *Condor*, **77**, 73–83.

Pankiw, T., Huang, Z.-Y., Winston, M. L., and Robinson, G. E. (1998). Queen mandibular gland pheromone influences worker honey bee (*Apis mellifera* L.) foraging ontogeny and juvenile hormone titers. *Journal of Insect Physiology*, **44**, 685–92.

Parker, G. A. and Hammerstein, P. (1985). Game theory and animal behaviour. In *Essays in honour of John Maynard Smith* (eds. P. J. Greenwood, P. H. Harvey, and M. Slatkin), pp. 73–94. Cambridge University Press, Cambridge.

Parr, L. A. and de Waal, F. B. M. (1999). Visual kin recognition in chimpanzees. *Nature*, **399**, 647–8.

Parrish, J. K. (1989). Re-examining the selfish herd: are central fish safer? *Animal Behaviour*, **38**, 1048–53.

Parrish, J. K. (1993). Comparisons of the hunting behavior of four piscine predators attacking schooling prey. *Ethology*, **95**, 233–46.

Parrish, J. K. and Edelstein-Keshet, L. (1999). Complexity, pattern, and evolutionary trade-offs in animal aggregation. *Science*, **284**, 99–101.

Parrish, J. K. and Hamner W. M. (1997). *Animal groups in three dimensions*. Cambridge University Press, Cambridge.

Parrish, J. K. and Turchin, P. (1997). Individual decisions, traffic rules, and emergent pattern in schooling fish. In *Animal groups in three dimensions* (eds. J. K. Parrish, and W. M. Hamner), pp. 126–42. Cambridge University Press, Cambridge.

Parrish, J. K., Strand, S. W., and Lott, J. L. (1989). Predation on a school of flatiron herring, *Harengual thrissina. Copeia*, **1989**, 1089–91.

Partridge, B. L. (1980). The effect of school size on the structure and dynamics of minnow schools. *Animal Behaviour*, **28**, 68–77.

Partridge, B. L. (1982). Structure and function of fish schools. *Scientific American*, **245**, 114–23.

Partridge, B. L. and Pitcher, T. J. (1979). Evidence against a hydrodynamic function for fish schools. *Nature*, **279**, 418–19.

Patterson, I. J. (1965). Timing and spacing of broods in the black-headed gull *Larus ridibundus*. *Ibis*, **107**, 433–59.

Patterson, M. R. (1984). Patterns of whole colony prey capture in the octocoral alcyonium siderium. *Biological Bulletin*, **167**, 613–19.

Pearce, J. M. (1997). *Animal learning and cognition*. Taylor & Francis Ltd., Psychology Press, Erlbaum, UK.

Pener, M. P. and Yerushalmi, Y. (1998). The physiology of locust phase polymorphism: an update. *Journal of Insect Physiology*, **44**, 365–77.

Peres, C. A. (1993). Anti-predator benefits in a mixed species group of amazonian tamarins. *Folia Primatologica*, **61**, 61–76.

Petit, D. R. and Bildstein, K. L. (1987). Effect of group size and location within the group on the foraging behavior of white ibises. *Condor*, **89**, 602–9.

Petrie, M. and Kempenaers, B. (1998). Extra-pair paternity in birds: explaining variation between species and populations. *Trends in Ecology and Evolution*, **13**, 52–8.

Petrie, M., Krupa, A., and Burke, T. (1999). Peacocks lek with relatives even in the absence of social and environmental cues. *Nature*, **401**, 155–7.

Pettifor, R. A. (1990). The effects of avian mobbing on a potential predator, the European kestrel, *Falco tinnunculus*. *Animal Behaviour*, **39**, 821–7.

Peuhkuri, N. (1997). Size-assortative shoaling in fish: the effect of oddity on foraging behaviour. *Animal Behaviour*, **54**, 271–8.

Peuhkuri, N. (1998). Shoal composition, body size and foraging in sticklebacks. *Behavioral Ecology and Sociobiology*, **43**, 333–7.

Peuhkuri, N. (1999). Size-assorted fish shoals and the majority's choice. *Behavioral Ecology and Sociobiology*, **46**, 307–12.

Peuhkuri, N., Ranta, E., and Seppä, P. (1997). Size-assortative schooling in free-ranging sticklebacks. *Ethology*, **103**, 318–24.

Phillips, A. V. and Stirling, I. (2000). Vocal individuality in mother and pup South American fur seals, *Arctocephalus australis*. *Marine Mammal Science*, **16**, 592–616.

Picman, J., Leonard, M., and Horn, A. (1988). Antipredation role of clumped nesting by marsh-nesting red-winged blackbirds. *Behavioral Ecology and Sociobiology*, **22**, 9–15.

Pitcher, T. J. (1983). Heuristic definitions of fish shoaling behavior. *Animal Behaviour*, **31**, 611–13.

Pitcher, T. J. and Parrish, J. K. (1993). Functions of shoaling behaviour in teleosts. In *Behaviour of teleost fishes* (ed. T. J. Pitcher), pp. 363–439. Chapman & Hall, London.

Pitcher, T. J., Magurran, A. E., and Winfield, I. J. (1982). Fish in larger shoals find food faster. *Behavioral Ecology and Sociobiology*, **10**, 149–51.

Pitcher, T. J., Magurran, A. E., and Allan, J. R. (1983). Shifts of behaviour with shoal size in cyprinids. *Proceedings of the 3rd British Freshwater Fisheries Conference*, **3**, 220–1.

Pitcher, T. J., Magurran, A. E., and Allan, J. R. (1986). Size segregative behaviour in minnow shoals. *Journal of Fish Biology, Supplement A*, **29**, 83–95.

Pius, S. M. and Leberg, P. L. (1997). Aggression and nest spacing in single and mixed species groups of seabirds. *Oecologia*, **111**, 144–50.

Poiani, A. (1992). Ectoparasitism as a possible cost of social life—a comparative analysis using Australian passerines (passeriforms). *Oecologica*, **92**, 429–41.

Poiani, A. and Yorke, M. (1989). Predator harassment: more evidence of deadly risk. *Ethology*, **83**, 167–9.

Porter, R. H. (1998). Olfaction and human kin recognition. *Genetica*, **104**, 259–63.

Post, W. and Seals, C. A. (1993). Nesting associations of least bitterns and boat-tailed grackles. *Condor*, **95**, 139–44.

Poulin, R. (1999). Parasitism and shoal size in juvenile sticklebacks: conflicting selection pressures from different ectoparasites. *Ethology*, **105**, 959–68.

Poulin, R. (2000). Manipulation of host behaviour by parasites: a weakening paradigm? *Proceedings of the Royal Society London Series B*, **267**, 787–92.

Poulin, R. F. (1991). Group-living and infestation by ectoparasites in passerines. *Condor*, **93**, 418–23.

Poulin, R. and FitzGerald, G. J. (1989). Shoaling as an anti-ectoparasite mechanism in juvenile sticklebacks (*Gasterosteus* spp.). *Behavioral Ecology and Sociobiology*, **24**, 251–5.

Pouyard, L., Desmarais, E., Chenuil, A., Agnese, J. F., and Bonhomme, F. (1999). Kin cohesiveness and possible inbreeding in the mouthbrooding tilapia *Sarotherodon melanotheron*. *Molecular Ecology*, **8**, 803–12.

Powell, G. V. N. (1974). Experimental analysis of the social value of flocking by starlings (*Sturnus vulgaris*) in relation to predation and foraging. *Animal Behaviour*, **22**, 501–5.

Pöysä, H. (1991). Does the attractiveness of teal foraging groups depend on the posture of group members? *Ornis Scandinavia*, **22**, 167–9.

Pöysä, H. (1992). Group foraging in patchy environments – the importance of coarse level local enhancement. *Ornis Scandinavica*, **23**, 159–66.

Price, G. R. (1970). Selection and covariance. *Nature*, **227**, 520–1.

Pritchard, V. L., Lawrence, J., Butlin, R. K., and Krause, J. (2001). Shoal size assessment in zebrafish, *Danio rerio*. *Animal Behaviour*, **62**, 1085–88.

Pruett-Jones, S. G. and Lewis, M. J. (1990). Sex ratio and habitat limitation promote delayed dispersal in superb fairy-wrens. *Nature*, **348**, 541–2.

Pugh, L. G. C. E. (1971). The influence of wind resistance in running and walking and the mechanical efficiency of work against horizontal or vertical forces. *Journal of Physiology*, **213**, 255–76.

Pulliam, H. R. (1973). On the advantages of flocking. *Journal of Theoretical Biology*, **38**, 419–22.

Pulliam, H. R., Pyke, G. H., and Caraco, T. (1982). The scanning behaviour of juncos: a game theoretical approach. *Journal of Theoretical Biology*, **95**, 89–103.

Putaala, A., Hohtola, E., and Hissa, R. (1995). The effect of group size on metabolism in huddling grey partridge (*Perdix perdix*). *Comparative Biochemistry and Physiology B—Biochemistry and Molecular Biology*, **111**, 243–7.

Quinn, T. P. and Busack, C. A. (1985). Chemosensory recognition of siblings in juvenile salmon (*Oncorhynchus kisutch*). *Animal Behaviour*, **33**, 51–6.

Quinn, T. P. and Hara, T. J. (1986). Sibling recognition and olfactory sensitivity in juvenile coho salmon (*Oncorhynchus kisutch*). *Canadian Journal of Zoology*, **64**, 921–5.

Radabaugh, D. C. (1980). Change in minnow, *Pimephales promelas* Rafineque, schooling behaviour associated with infections of brain-encysted larvae of the fluke, *Ornithodiplostomum ptychocheilus*. *Journal of Fish Biology*, **16**, 621–8.

Randerson, J. P., Jiggins, F. M., and Hurst, L. D. (2000). Male killing can select for male mate choice: a novel solution to the paradox of the lek. *Proceedings of the Royal Society London Series B*, **267**, 867–74.

Ranta, E. and Lindström, K. (1990). Assortative schooling in three-spined sticklebacks? *Annales Zoologici Fennici*, **27**, 67–75.

Ranta, E., Lindström, K., and Peuhkuri, N. (1992). Size matters when three-spined stickle-backs go to school. *Animal Behaviour*, **43**, 160–2.

Ranta, E., Rita, H., and Lindström, K. (1993). Competition versus cooperation—success of individuals foraging alone and in groups. *American Naturalist*, **142**, 42–58.

Ranta, E., Peuhkuri, N., and Laurila, A. (1994). A theoretical exploration of antipredatory and foraging factors promoting phenotype-assorted fish schools. *Ecoscience*, **1**, 99–106.

Ratchford, S. G. and Eggleston, D. B. (1998). Size- and scale-dependent chemical attraction contribute to an ontogenetic shift in sociality. *Animal Behaviour*, **56**, 1027–34.

Rattenborg, N. C., Lima, S. L., and Amlaner, C. J. (1999). Facultative control of avian uni-hemispheric sleep under the risk of predation. *Behavioural Brain Research*, **105**, 163–72.

Rautio, S. A., Bura, E. A., Berven, K. A., and Gamboa, G. J. (1991). Kin recognition in wood frog tadpoles (*Rana sylvatica*)—factors affecting spatial proximity to siblings. *Canadian Journal of Zoology*, **69**, 2569–71.

Rayor, L. S. and Uetz, G. W. (1990). Trade-offs in foraging success and predation risk with spatial position in colonial spiders. *Behavioral Ecology and Sociobiology*, **27**, 77–85.

Rayor, L. S. and Uetz, G. W. (1993). Ontogenetic shifts within the selfish herd: predation risk and foraging trade-offs change with age in colonial web-building spiders. *Oecologia*, **95**, 1–8.

Reader, S. M. and Laland, K. N. (2000). Diffusion of foraging innovations in the guppy. *Animal Behaviour*, **60**, 175–80.

Reebs, S. G. (2000). Can a minority of informed leaders determine the foraging movements of a fish shoal? *Animal Behaviour*, **59**, 403–9.

Reebs, S. G. (2001). Influence of body size on leadership in shoals of golden shiners, *Notemigonus crysoleucas*. *Behaviour*, **138**, 797–809.

Reebs, S. G. and Gallant, B. Y. (1997). Food-anticipatory activity as a cue for local enhance-ment in golden shiners. *Ethology*, **103**, 1060–9.

Reebs, S. G. and Saulnier, N. (1997). The effect of hunger on shoal choice of golden shiners (Pisces: Cyprinidae, *Notemigonus crysoleucas*). *Ethology*, **103**, 642–52.

Reeve, H. K. and Keller, L. (1999). Levels of selection: burying the units-of-selection debate and unearthing the crucial new issues. In *Levels of selection* (ed. L. Keller). Princeton University Press, Princeton.

Reeve, H. K. and Ratnieks, F. L. W. (1993). Resolutions of conflicts in polygynous societies: mutual tolerance and reproductive skew. In *Queen number and sociality in insects* (ed. L. Keller), pp. 45–85. Oxford University Press, Oxford.

Reuter, H. and Breckling, B. (1994). Selforganization of fish schools—an object-oriented model. *Ecological Modelling*, **75**, 147–59.

Reznick, D. N., Bryga, H., and Endler, J. A. (1990). Experimentally induced life-history evolution in a natural population. *Nature*, **346**, 357–9.

Richner, H. and Heeb, P. (1995). Is the information centre hypothesis a flop? *Advances in the Study of Behavior*, **24**, 1–45.

Richner, H. and Heeb, P. (1996). Communal life: honest signaling and the recruitment centre hypothesis. *Behavioral Ecology*, **7**, 115–19.

Riechert, S. E. and Hedrick, A. V. (1990). Levels of predation and genetically based anti-predator behaviour in the spider, *Agelenopsis aperta*. *Animal Behaviour*, **40**, 679–87.

Riechert, S. E. and Roeloffs, R. M. (1993). Evidence for and consequences of inbreeding in the cooperative spiders. In *The natural history of inbreeding and outbreeding* (ed. N. Thornhill), pp. 283–303. The University of Chicago Press, Chicago and London.

Riedman, M. L. (1982). The evolution of alloparental care and adoption in mammals and birds. *Quarterly Review of Biology*, **57**, 405–35.

Riipi, M., Alatalo, R. V., Lindstrom, L., and Mappes, J. (2001). Multiple benefits of gregariousness cover detectability costs in aposematic aggregations. *Nature*, **413**, 512–14.

Ritz, D. A. (2000). Is social aggregation in aquatic crustaceans a strategy to conserve energy? *Canadian Journal of Aquatic Science*, **57**, 59–67.

Roberts, G. (1995). A real-time response of vigilance behaviour to changes in group size. *Animal Behaviour*, **50**, 1371–4.

Roberts, G. (1996). Why vigilance declines as group size increases. *Animal Behaviour*, **51**, 1077–86.

Robinson, S. K. (1985). Coloniality in the yellow-rumped cacique as a defense against predators. *Auk*, **102**, 506–19.

Robinson, G. E. (1992). Regulation of division of labor in insect societies. *Annual Review of Entomology*, **37**, 637–65.

Robinson, C. M. and Pitcher, T. J. (1989). The influence of hunger and ration level on shoal density, polarisation and swimming speed of herring, *Clupea harengus* L. *Journal of Fish Biology*, **34**, 631–3.

Rodman, P. S. (1981). Inclusive fitness and group size with a reconsideration of group sizes in lions and wolves. *American Naturalist*, **118**, 275–83.

Röell, A. (1978). Social behaviour of the jackdaw, *Corvus monedula*, in relation to its niche. *Behaviour*, **64**, 1–124.

Roessingh, P., Simpson, S. J., and James, S. (1993). Analysis of phase-related changes in the behaviour of desert locust nymphs. *Proceedings of the Royal Society London Series B*, **252**, 43–9.

Romey, W. L. (1995). Position preferences within groups: do whirligigs select positions which balance feeding opportunities with predator avoidance? *Behavioral Ecology and Sociobiology*, **37**, 195–200.

Romey, W. L. (1996). Individual differences make a difference in the trajectory of simulated schools of fish. *Ecological Modelling*, **92**, 65–77.

Romey, W. L. (1997). Inside or outside? Testing evolutionary predictions or positional effects. In *Animal groups in three dimensions* (eds. J. K. Parrish and W. M. Hamner), pp. 174–93. Cambridge University Press, Cambridge.

Rozsa, L., Rekasi, J., and Reiczigel, J. (1996). Relationship of host coloniality to the population ecology of avian lice (Insecta: Phthiraptera). *Journal of Animal Ecology*, **65**, 242–8.

Rubenstein, D. I. and Hohmann, M. E. (1989). Parasites and social behaviour of island feral horses. *Oikos*, **55**, 312–20.

Rutberg, A. T. (1987). Horse fly harassment and the social behaviour of feral ponies. *Ethology*, **75**, 145–54.

Ruxton, G. D. (1993). Foraging in flocks—nonspatial models may neglect important costs. *Ecological Modelling*, **82**, 277–85.

Ruxton, G. D., Gurney, W. S. C., and de Roos, A. M. (1992). Interference and generation cycles. *Theoretical Population Biology*, **42**, 235–53.

Ruxton, G. D., Hall, S. J., and Gurney, W. S. C. (1995). Attraction towards feeding conspecifics when food patches are exhaustible. *American Naturalist*, **145**, 653–60.

Ruzzante, D. E. and Doyle, R. W. (1991). Rapid behavioral changes in medaka (*Oryzias latipes*) caused by selection for competitive and noncompetitive growth. *Evolution*, **45**, 1936–46.

Ruzzante, D. E. and Doyle, R. W. (1993). Evolution of social behaviour in a resource-rich, structured environment: selection experiments with medaka (*Oryzias latipes*). *Evolution*, **47**, 456–70.

Ryer, C. H. and Olla, B. L. (1991). Information transfer and the facilitation and inhibition of feeding in schooling fish. *Environmental Biology of Fishes*, **30**, 317–23.

Sakakura, Y. and Tsukamoto, K. (1996). Onset and development of cannibalistic behaviour in early life stages of yellowtail. *Journal of Fish Biology*, **48**, 16–29.

Sasvari, L. (1992). Great tits benefit from feeding in mixed-species flocks: a field study. *Animal Behaviour*, **43**, 289–96.

Schaller, G. B. (1972). *The Serengeti Lion*. University of Chicago Press, Chicago.

Schieck, J. O. and Hannon, S. J. (1993). Clutch predation, cover, and the overdispersion of nests of the willow ptarmigan. *Ecology*, **74**, 743–50.

Schmidt, P. A. and Mech, L. D. (1997). Wolf pack size and food acquisition. *American Naturalist*, **150**, 513–17.

Schmitt, R. J. and Strand, S. W. (1982). Cooperative foraging by yellowtail, *Seriola lalandei* (Carangidae) on two species of fish prey. *Copeia*, **1982**, 714–17.

Schradin, C. (2000). Confusion effect in a reptilian and a primate predator. *Ethology*, **106**, 691–700.

Schulz, D. J. and Robinson, G. E. (2001). Octopamine influences division of labor in honey bee colonies. *Journal of Comparative Physiology A*, **187**, 53–61.

Schulz, D. J., Huang, Z. Y., and Robinson, G. E. (1998). Effects of colony food shortage on behavioral development in honey bees. *Behavioral Ecology and Sociobiology*, **42**, 295–303.

Segers, J. (1983). Partial bivoltinism may cause alternating sex-ratio biases that favour eusociality. *Nature*, **301**, 59–62.

Segers, J. (1991). Cooperation and conflict in social insects. In *Behavioural ecology: an evolutionary approach* (eds. J. R. Krebs and N. B. Davies), pp. 338–73, 3rd edn. Blackwell Scientific Publications, Oxford.

Seghers, B. H. (1974). Schooling behaviour in the guppy: an evolutionary response to predation. *Evolution*, **28**, 486–9.

Seghers, B. H. (1981). Facultative schooling behavior in the spottail shiner (*Notropis hudsonius*): possible costs and benefits. *Environmental Biology of Fishes*, **6**, 21–4.

Selander, R. K. (1964). Speciation in wrens of the genus *Campylorhynchus*. *University of California Publications in Zoology*, **74**, 1–224.

Selman, J. and Goss-Custard, J. D. (1988). Interference between foraging redshank, *Tringa totanus*. *Animal Behaviour*, **36**, 1542–4.

Sendova-Franks, A. B. and Franks, N. R. (1998). Self-assembly, self-organization and division of labour. *Proceedings of the Royal Society London Series B*, **354**, 1395–405.

Shaw, P. W., Carvalho, G. R., Magurran, A. E., and Seghers, B. H. (1991). Population differentiation in Trinidadian guppies (*Poecilia reticulata*): patterns and problems. *Journal of Fish Biology*, **39**, 203–9.

Shaw, P. W., Carvalho, G. R., Magurran, A. E., and Seghers, B. H. (1994). Factors affecting the distribution of genetic variability in a freshwater fish, the guppy *Poecilia reticulata*. *Journal of Fish Biology*, **45**, 875–88.

Shorey, L., Piertney, S., Stone, J., and Höglund, J. (2000). Fine-scale genetic structuring on *Manacus manacus* leks. *Nature*, **408**, 352–3.

Sibly, R. M. (1983). Optimal group size is unstable. *Animal Behaviour*, **31**, 947–8.

Siegfried, W. R. and Underhill, L. G. (1975). Flocking as an anti-predator strategy in doves. *Animal Behaviour*, **23**, 504–8.

Simpson, S. J., McCaffery, A. R., and Hägele, B. F. (1999). A behavioural analysis of phase change in the desert locust. *Biological Reviews*, **74**, 461–80.

Simpson, S. J., Despland, E., Hägele, B. F., and Dodgson, T. (2001). Gregarious behavior in desert locusts is evoked by touching their back legs. *Proceedings of the National Academy of Sciences of the USA*, **98**, 3895–7.

Sinclair, A. R. E. (1977). *The African Buffalo*. University of Chicago Press, Chicago.

Sinclair, A. R. E. (1985). Does interspecific competition or predation shape the African ungulate community? *Journal of Animal Ecology*, **54**, 899–918.

Sirot, E. (2000). An evolutionarily stable strategy for aggressiveness in feeding groups. *Behavioral Ecology*, **11**, 351–6.

Smith, R. T. (1978). Roosting of long-tailed tits. *British Birds*, **71**, 362.

Smith, E. A. (1981). The application of optimal foraging theory to the analysis of hunter-gatherer group size. In *Hunter-gatherer foraging strategies* (eds. B. Winterhalder and E. A. Smith), pp. 36–65. Chicago University Press, Chicago.

Smith, J. W., Benkman, C. W., and Coffey, K. (1999). The use and misuse of public information by foraging red crossbills. *Behavioural Ecology*, **10**, 54–62.

Sober, E. and Wilson, D. S. (1998). *Unto others: the evolution and psychology of unselfish behaviour*. Harvard University Press, Cambridge, MA.

Sonerud, G. A., Smedshaug, C. A., and Bråthen, Ø. (2001). Ignorant hooded crows follow knowledgeable roost-mates to food: support for the information centre hypothesis. *Proceedings of the Royal Society London Series B*, **268**, 827–31.

Sorci, G., deFraipont, M., and Clobert, J. (1997). Host density and ectoparasite avoidance in the common lizard (*Lacerta vivipara*). *Oecologica*, **111**, 183–8.

Speakman, J. R. and Banks, D. (1998). The function of flight formations in Greylag Geese *Anser anser*; energy saving or orientation? *Ibis*, **140**, 280–7.

Stander, P. E. (1992). Cooperative hunting in lions: the role of the individual. *Behavioral Ecology and Sociobiology*, **29**, 445–54.

Steck, N., Wedekind, C., and Milinski, M. (1999). No sibling odor preference in juvenile three-spined sticklebacks. *Behavioral Ecology*, **10**, 493–7.

Stillman, R. A., Goss-Custard, J. D., and Caldow, R. W. G. (1997). Modelling interference from basic foraging behaviour. *Journal of Animal Ecology*, **66**, 692–703.

Stillman, R. A., Goss-Custard, J. D., Clarke, R. T., and Dit Durell, S. E. A. Le V. (1996). Shape of the interference function in a foraging vertebrate. *Journal of Animal Ecology*, **65**, 813–24.

Stillman, R. A., Goss-Custard, J. D., and Alexander, M. J. (2000). Predator search pattern and the strength of interference through prey depression. *Behavioral Ecology*, **11**, 597–605.

Stookey, J. M. and Gonyou, H. W. (1998). Recognition in swine: recognition through familiarity or genetic relatedness? *Applied Animal Behaviour Science*, **55**, 291–305.

Sugden, L. G. and Betersbergen, G. W. (1986). Effect of density and concealment on American Crow predation of simulated duck nests. *Journal of Wildlife Management*, **50**, 9–14.

Sullivan, J. P., Jassim, O., Fahrbach, S. E., and Robinson, G. E. (2000). Juvenile hormone paces behavioral development in the adult worker honey bee. *Hormones and Behavior*, **37**, 1–14.

Sutherland, W. J. (1996). *From individual behaviour to population ecology*. Oxford University Press, Oxford.

Svensson, P. A., Barber, I., and Forsgren, E. (2000). Shoaling behaviour of the two-spotted goby. *Journal of Fish Biology*, **56**, 1477–87.

Sweeney, B. W. and Vannote, R. L. (1982). Population synchrony in mayflies: a predator satiation hypothesis. *Evolution*, **36**, 810–21.

Sword, G. A., Simpson, S. J., El Hadi, O. T. M., and Wilps, H. (2000). Density-dependent aposematism in the desert locust. *Proceedings of the Royal Society London Series B*, **267**, 63–68.

Szép, T. and Barta, Z. (1992). The threat to bank swallows from the hobby at a large colony. *Condor*, **94**, 1022–5.

Taylor, R. H. (1962). The Adelie penguin *Pygoscelis adeliae* at Cape Royds. *Ibis*, **104**, 176–204.

Tegeder, R. W. and Krause, J. (1995). Density dependence and numerosity in fright stimulated aggregation behaviour of shoaling fish. *Philosophical Transactions of the Royal Society of London B*, **350**, 381–90.

Templeton, J. J. and Giraldeau, L.-A. (1995a). Public information cues affect the scrounging decisions of starlings. *Animal Behaviour*, **49**, 1617–26.

Templeton, J. J. and Giraldeau, L.-A. (1995b). Patch assessment in foraging flocks of European starlings—evidence for the use of public information. *Behavioral Ecology*, **6**, 65–72.

Templeton, J. J. and Giraldeau, L.-A. (1996). Vicarious sampling: the use of personal and public information by starlings foraging in a simple patchy environment. *Behavioral Ecology and Sociobiology*, **38**, 105–14.

Theodorakis, C. W. (1989). Size segregation and the effects of oddity on predation risk in minnow shoals. *Animal Behaviour*, **38**, 496–502.

Thorpe, W. H. (1958). The learning of song patterns by birds, with especial reference to the song of the chaffinch *Fringilla coelebs*. *Ibis*, **100**, 535–70.

Tinbergen, N. (1951). On the significance of territory in the herring gull. *Ibis*, **94**, 158–9.

Tinbergen, N. (1963). On aims and methods of ethology. *Zeitschrift für Tierpsychologie*, **20**, 410–33.

Tinbergen, N., Impekoven, M., and Franck, D. (1967). An experiment on spacing-out as a defense against predation. *Behaviour*, **28**, 307–21.

Towers, S. R. and Coss, R. G. (1990). Confronting snakes in the burrow: snake-species discrimination and anti-snake tactics of two California ground squirrel populations. *Ethology*, **84**, 177–92.

Trail, P. W. (1987). Predation and antipredator behaviour at Guianan cock-of-the-rock leks. *Auk*, **104**, 496–507.

Treherne, J. E. and Foster, W. A. (1980). The effects of group size on predator avoidance in a marine insect. *Animal Behaviour*, **28**, 1119–22.

Treherne, J. E. and Foster, W. A. (1981). Group transmission of predator avoidance behaviour in a marine insect: the Trafalgar effect. *Animal Behaviour*, **29**, 911–17.

Treherne, J. E. and Foster, W. A. (1982). Group size and anti-predator strategies in a marine insect. *Animal Behaviour*, **32**, 536–42.

Treisman, M. (1975). Predation and the evolution of gregariousness. I. Models for concealment and evasion. *Animal Behaviour*, **23**, 779–800.

Trivers, R. L. (1971). The evolution of reciprocal altruism. *Quarterly Reviews of Biology*, **46**, 35–57.

Trivers, R. L. and Hare, H. (1976). Haplodiploidy and the evolution of social insects. *Science*, **191**, 249–63.

Tulley, J. J. and Huntingford, F. A. (1987a). Age, experience and the development of adaptive variation in anti-predator responses in three-spined sticklebacks (*Gasterosteus aculeatus*). *Ethology*, **75**, 285–90.

Tulley, J. J. and Huntingford, F. A. (1987b). Parental care and the development of adaptive variation in anti-predator responses in sticklebacks. *Animal Behaviour*, **35**, 1570–2.

Turner, G. F. and Pitcher T. J. (1986). Attack abatement: a model for group protection by combined avoidance and dilution. *American Naturalist*, **128**, 228–40.

Valone, T. J. (1989). Group foraging, public information and patch estimation. *Oikos*, **56**, 357–63.

Valone, T. J. (1993). Patch information estimation—a cost of group foraging. *Oikos*, **68**, 258–66.

Van Havre, N. and FitzGerald, G. J. (1988). Shoaling and kin recognition in the threespine stickleback (*Gasterosteus aculeatus* L.). *Biology of Behaviour*, **13**, 190–201.

Van Kampen, H. S. (1994). Courtship food-calling in Burmese red junglefowl. 1. The causation of female approach. *Behaviour*, **131**, 261–75.

Van Kampen, H. S. (1997). Courtship food-calling in Burmese red junglefowl. 2. Sexual conditioning and the role of the female. *Behaviour*, **134**, 775–87.

Van Kampen, H. S. and Hogan, J. A. (2000). Courtship food-calling in Burmese red junglefowl. III. Factors influencing the male's behaviour. *Behaviour*, **137**, 1191–209.

Van Schaik, C. P., van Noordwijk, M. A., Warsono, B., and Sutriono, E. (1983). Party size and early detection of predators in Sumatran forest primates. *Primates*, **24**, 211–12.

Van Vuren, D. (1996). Ectoparasites, fitness, and social behaviour in yellow-bellied marmots. *Ethology*, **102**, 686–94.

Veen, J. (1977). Functional and causal aspects of nest distribution in colonies of the sandwich tern (*Sterna S. sandvicensis*). *Behaviour Supplement*, **20**, 1–193.

Vehrencamp, S. L. (1983a). Optimal degree of skew in cooperative societies. *American Zoologist*, **23**, 327–35.

Vehrencamp, S. L. (1983b). A model for the evolution of despotic versus egalitarian societies. *Animal Behaviour*, **31**, 667–82.

Vickery, W. L. and Millar, J. S. (1984). The energetics of huddling by endotherms. *Oikos*, **43**, 88–93.

Videler, J. J. (1993). *Fish swimming*. Chapman & Hall, London.

Vine, I. (1973). Detection of prey flocks by predators. *Journal of Theoretical Biology*, **40**, 207–10.

Vollrath, F. (1986). Eusociality and extraordinary sex ratios in the spider *Anelosimus eximius* (Araneae: Theridiidae). *Behavioral Ecology and Sociobiology*, **18**, 283–7.

Wade, M. J. (1976). Group selection among laboratory populations of Tribolium. *Proceedings of the National Academy of Sciences of the USA*, **73**, 4604–7.

Wagner, R. H., Danchin, E., Boulinier, T., and Helfenstein, F. (2000). Colonies as byproducts of commodity selection. *Behavioral Ecology*, **11**, 572–3.

Waldman, B. (1984). Kin recognition and sibling association among wood frog (*Rana sylvatica*) tadpoles. *Behavioral Ecology and Sociobiology*, **14**, 171–80.

Warburton, K. (1997). Social forces in animal congregations: interactive, motivational, and sensory aspects. In *Animal groups in three dimensions* (eds. J. K. Parrish and W. M. Hamner), pp. 313–36. Cambridge University Press, Cambridge.

Warburton, K. and Lees, N. (1996). Species discrimination in guppies: learned response to visual cues. *Animal Behaviour*, **52**, 371–8.

Ward, A. J. W. and Krause, J. (2001). Body length assortative shoaling in the European minnow. *Animal Behaviour*, **62**, 617–21.

Ward, A. J. W., Hoare, D. J., Couzin, I. D., Broom, M., and Krause, J. (2002). The effects of parasitism and body length on the positioning within wild fish shoals. *Journal of Animal Ecology*, **71**, 10–14.

Ward, P. and Zahavi, A. (1973). The importance of certain assemblages of birds as "information centres" for food finding. *Ibis*, **115**, 517–34.

Watkins, J. L., Buchholz, F., Priddle, J., Morris, D. J., and Ricletts, C. (1992). Variation in reproductive status of Antarctic krill swarms; evidence for a size-related sorting mechanism? *Marine Ecology Progress Series*, **82**, 163–74.

Watt, P. J., Nottingham, S. F., and Young, S. (1997). Toad tadpole aggregation behaviour: evidence for a predator avoidance function. *Animal Behaviour*, **54**, 865–72.

Weihs, D. (1973). Hydrodynamics of fish schooling. *Nature*, **241**, 290–1.

Weihs, D. (1975). Some hydrodynamical aspects of fish schooling. In *Symposium on swimming and flying in nature* (eds. T. Y. Wu, C. J. Brokaw, and C. Brennen). Plenum Press, New York.

Weimerskirch, H., Martin, J., Clerquin, Y., Alexandre, P., and Jiraskova, S. (2001). Energy saving in flight formation. *Nature*, **413**, 697–8.

Weisser, W. W., Wilson, H. B., and Hassell, M. P. (1997). Interference among parasitoids: a clarifying note. *Oikos*, **79**, 173–8.

Westneat, D. F. and Sherman, P. W. (1997). Density and extra-pair fertilisations in birds: a comparative analysis. *Behavioral Ecology and Sociobiology*, **41**, 205–15.

Westneat, D. F., Walters, A., McCarthy, T. M., Hatch, M. I., and Hein, W. K. (2000). Alternative mechanisms of non-independent mate choice. *Animal Behaviour*, **59**, 467–76.

Wetterer, J. K. and Bishop, C. J. (1985). Planktivore prey selection: the reactive field volume vs. the apparent size model. *Ecology*, **66**, 457–64.

Wheelwright, N. T., Lawler, J. J., and Weinstein, J. H. (1997). Nest-site selection in Savannah sparrows: using gulls as scarecrows. *Animal Behaviour*, **53**, 197–208.

Widemo, F. and Owens, I. P. F. (1995). Lek size, male mating skew and the evolution of lekking. *Nature*, **373**, 148–50.

Wieser, W., Forstner, H., Medgyesy, N., and Hinterleitner, S. (1988). To switch or not to switch: partitioning of energy between growth and activity in larval cyprinids (Cyprinidae: Teleostei). *Functional Ecology*, **2**, 499–507.

Wiley, R. H. (1991). Lekking in birds and mammals: behavioural and evolutionary issues. *Advances in the Study of Behavior*, **20**, 201–91.

Wiley, R. H. and Rabenold, K. N. (1984). The evolution of cooperative breeding by delayed reciprocity and queuing for favourable social positions. *Evolution*, **38**, 609–21.

Wilkinson, G. S. (1984). Reciprocal food sharing in the vampire bat. *Nature*, **308**, 181–4.

Wilkinson, G. S. and Boughman, J. W. (1998). Social calls coordinate foraging in greater spear-nosed bats. *Animal Behaviour*, **55**, 337–50.

Williams, G. C. (1966). *Adaptation and natural selection*. Princeton University Press, Princeton.

Wilson, D. S. (1974). Prey capture and competition in the ant lion. *Biotropica*, **6**, 187–93.

Wilson, D. S. (1980). *The natural selection of populations and communities*. Benjamin Cummings, Menlo Park, California.

Wilson, D. S. (1997). Evolution—human groups as units of selection. *Science*, **276**, 1816–17.

Wilson, O. E. (1975). *Sociobiology, the modern synthesis*. Harvard University Press, Cambridge.

Winberg, S. and Olsén, K. H. (1992). The influence of rearing conditions on the sibling odor preference of juvenile arctic charr, *Salvelinus alpinus* L. *Animal Behaviour*, **44**, 157–64.

Wirtz, P. and Lörscher, J. (1983). Group sizes of antelopes in an East African park. *Behaviour*, **84**, 135–55.

Witte, K. and Ryan, M. J. (1998). Male body length influences mate-choice copying in the sailfin molly *Poecilia latipinna*. *Behavioral Ecology*, **9**, 534–9.

Wittenberger, J. F. and Hunt, G. L. Jr. (1985). The adaptive significance of coloniality in birds. In *Avian biology Volume 8* (eds. D. S. Farner, J. R. King, and K. C. Parkes), pp. 1–78. Academic Press, New York.

Wolf, N. (1985). Odd fish abandon mixed-species groups when threatened. *Behavioral Ecology and Sociobiology*, **17**, 47–52.

Woolfenden, G. E. and Fitzpatrick, J. W. (1978). The inheritance of territory in group breeding birds. *BioScience*, **28**, 104–8.

Wrona, F. and Dixon, R. W. J. (1991). Group size and predation risk: a field analysis of encounter and dilution effects. *American Naturalist*, **137**, 186–201.

Wynne-Edwards, V. C. (1962). *Animal dispersal in relation to social behaviour*. Oliver & Boyd, Edinburgh.

Wynne-Edwards, V. C. (1993). A rationale for group selection. *Journal of Theoretical Biology*, **162**, 1–22.

Ydenberg, R. C. and Dill, L. M. (1986). The economics of fleeing from predators. *Advances in the Study of Behavior*, **16**, 229–49.

Yom-Tov, Y. (2001). An updated list and some comments on the occurrence of intraspecific nest parasitism in birds. *Ibis*, **143**, 133–43.

Zahavi, A. (1977). Reliability in communication systems and the evolution of altruism. In *Evolutionary ecology* (eds. B. Stonehouse and C. M. Perrins), pp. 253–9. Macmillan, London.

Zemel, A. and Lubin, Y. (1995). Inter-group competition and stable group sizes. *Animal Behaviour*, **50**, 485–8.

Zuyev, G. V. and Belyayev, V. V. (1970). An experimental study of the swimming of fish in groups as exemplified by the horsemackerel. *Journal of Ichthyology*, **10**, 545–9.

Author Index

General Index